Science and Mythology in the
Making of Defence Policy

Also available from Brassey's

CHARTERS & TUGWELL
Armies in Low-intensity Conflict

CIMBALA
Ending a Nuclear War

DOUGLASS
Why the Soviets Violate Arms Control Treaties

GREENE & STROM
"Would the Insects Inherit the Earth"

GROVE & WINDASS
The Crucible of Peace

HARTCUP
The War of Invention: Scientific Developments 1914–18

LIN
New Weapon Technologies & The ABM Treaty

RUSI/BRASSEY'S
RUSI/Brassey's Defence Yearbook 1989

SHEARMAN & WILLIAMS
The Superpowers, Central America and the Middle East

TAYLOR
The Terrorist

Science and Mythology in the Making of Defence Policy

Edited by
Margaret Blunden
and
Owen Greene

BRASSEY'S DEFENCE PUBLISHERS
(a member of the Maxwell Pergamon Publishing Corporation plc)
LONDON · OXFORD · WASHINGTON · NEW YORK
BEIJING · FRANKFURT · SÃO PAULO · SYDNEY · TOKYO · TORONTO

U.K. (Editorial)	Brassey's Defence Publishers Ltd., 24 Gray's Inn Road, London WC1X 8HR
(Orders)	Brassey's Defence Publishers Ltd., Headington Hill Hall, Oxford OX3 0BW, England
U.S.A. (Editorial)	Pergamon-Brassey's International Defense Publishers, Inc., 8000 Westpark Drive, Fourth Floor, McLean, Virginia 22102, U.S.A.
(Orders)	Pergamon Press, Inc., Maxwell House, Fairview Park, Elmsford, New York 10523, U.S.A.
PEOPLE'S REPUBLIC OF CHINA	Pergamon Press, Room 4037, Qianmen Hotel, Beijing, People's Republic of China
FEDERAL REPUBLIC OF GERMANY	Pergamon Press GmbH, Hammerweg 6, D-6242 Kronberg, Federal Republic of Germany
BRAZIL	Pergamon Editora Ltda, Rua Eça de Queiros, 346, CEP 04011, Paraiso, São Paulo, Brazil
AUSTRALIA	Pergamon-Brassey's Defence Publishers Pty Ltd., P.O. Box 544, Potts Point, N.S.W. 2011, Australia
JAPAN	Pergamon Press, 5th Floor, Matsuoka Central Building, 1-7-1 Nishishinjuku, Shinjuku-ku, Tokyo 160, Japan
CANADA	Pergamon Press Canada Ltd., Suite No 271, 253 College Street, Toronto, Ontario, Canada M5T 1R5

Copyright © 1989 Brassey's Defence Publishers Ltd.

All Rights Reserved. No part of this publication may be reproduced, stored in a retrieval system or transmitted in any form or by any means: electronic, electrostatic, magnetic tape, mechanical, photocopying, recording or otherwise, without permission in writing from the publishers.

First edition 1989

Library of Congress Cataloging in Publication Data
Science and mythology in the making of defence
policy/edited by Margaret Blunden and Owen Greene.
p. cm.
1. Military policy—Decision making. 2. Great Britain—
Military—Decision making.
I. Blunden, Margaret. II. Greene, Owen.
UA11.S345 1988   355'.0335—dc 19   88-19818

British Library Cataloguing in Publication Data
Science and mythology in the making of
defence policy.
1. Western world. Defence. Policies of government
I. Blunden, Margaret, *1939–* II. Greene, Owen
335'.0335'1821

ISBN 0-08-033621-3

Printed in Great Britain by A. Wheaton & Co. Ltd., Exeter

Contents

Acknowledgements		vi

Part I

	Introduction: Science and the Defence Policy-making Process	3
1.	The Risk of Nuclear Winter: Scientific Research and Policy Debate OWEN GREENE AND JOYCE TAIT	13
2.	Rendering Nuclear Weapons Impotent and Obsolete: the Origins of a Pipedream DAVID CARLTON	61
3.	The Alchemists of Our Time: the Weapons Scientist as Scapegoat MARGARET BLUNDEN, OWEN GREENE AND JOHN NAUGHTON	77

Part II

	Introduction: Civil–Military Interactions: Fragmentation, Distortion and Policy Making	121
4.	Nuclear Weapons and Nuclear Power: Bias and Mythology in the Making of the British Magnox Nuclear Reactor Programme DAVID LOWRY	127
5.	Policy Making in Civil and Military Electronics: the Limits of Pragmatism MARGARET BLUNDEN, CHRIS BISSELL AND JOHN MONK	167
6.	British Defence Decision Making: the Boundaries of Influence MARGARET BLUNDEN	205
	About the Authors	246
	Index	247

Acknowledgements

A large number of people have been generous with their help in the preparation of this book. The editors would particularly like to thank Janet Radcliffe Richards, John Simpson, Patricia Lewis, Sue Willett, Phil Gummett, Steve Scofield, Peter Southwood, Gloria Franklin, Laurence Freedman, Laurence Martin, Sir James Eberle, David Bolton, Sir Frank Cooper and Sir John Nott.

Grateful acknowledgement is made to the following: to the Jane's Publishing Company Ltd. for permission to quote from *Jane's Fighting Ships, 1981–2*; to Oxford University Press for permission to quote from *U.K. Military R and D*; to the Armament and Disarmament Information Unit, University of Sussex, for permission to reproduce two figures from ADIU Report volume 8; to Lloyds Bank plc for permission to quote from the Lloyds Bank Review of October 1986; and to Her Majesty's Stationery Office for permission to quote from Statements on the Defence Estimates, from Reports of the House of Commons Defence Committee and from Hansard.

The editors are also grateful to the authors for their willingness to rethink and revise their contributions during the iterative and highly collaborative evolution of this book. The responsibility for each chapter is of course that of the authors alone.

PART I

INTRODUCTION
Science and the Defence Policy-making Process

In a memorandum entitled *Scientific and Technological Resources as Military Assets*, General Eisenhower told the United States Army in 1946 that 'The future security of the nation demands that all those civilian resources which by conversion or redirection constitute our main support in times of emergency be associated closely with the activities of the Army in time of peace' and advised extensive contracting for scientific and industrial services.[1]

Since that time, the expansion of the role of science and technology in defence has seemed to be irresistible. Massive research and development has structured the options for the future. Science and scientists have acquired an immense prestige. Moreover, the scale and complexity of the tasks involved in defence planning are such that 'scientific' approaches to management are the order of the day. 'Expert' reports and commissions are called upon to illuminate the implications of various trends and policy options. No policy position is quite respectable unless it draws on scientific prestige or expert knowledge.

Nevertheless, defence debates and policy making remain endemically incoherent and confused. Different experts endorse mutually incompatible positions. Scientific studies are commissioned or interpreted according to lobbyists' requirements. Science, technology and the production of 'knowledge' are all to some extent shaped by the policy-making process.

These observations will hardly surprise people who have studied or participated in defence decision-making. Nevertheless, myths about the relationship between science, experts and policy are widespread. Indeed, politicians, officials and lobbyists often have an interest in reinforcing them. The chapters of this book are all concerned, in their different ways, with exploring and illuminating these relationships and, in the process, with debunking some of the convenient myths about them.

The dominant notion of the role of science and technology in defence policy is of the obedient but extraordinarily capable servant of a national security policy defined by government and the military. This reflects the formal position of research and development (R & D) in the decision-making hierarchy and also bolsters military and political leaders' sense of their own power and authority. For example, it permeates President Reagan's celebrated television address on the Strategic Defense Initiative (SDI) of 23 March 1983:

> I call upon the scientific community in our country, those who gave us nuclear weapons, to turn their great talents now to the cause of mankind and world peace, to give us the means of rendering these nuclear weapons impotent and obsolete.[2]

The SDI programme took most people, including many senior defence officials within the Reagan administration itself, by surprise. A heavily funded R & D programme was established in order to pursue an objective formulated by the highest political authorities — to develop the technological capacity for a leak-proof shield against missile attack. However, since the feasibility of Reagan's vision relies so heavily on novel technologies, some notion of the capacity of scientists to meet the challenges posed by the SDI was clearly important to the formulation of the programme.

The role of scientific and technological advice in Reagan's vision of rendering nuclear missiles obsolete is explored in David Carlton's chapter in this book. However, it is worthwhile raising some of the issues at this stage since the SDI programme touches on themes that are more broadly relevant.

In the early 1980s, American military R & D establishments were increasingly engaged in space technology projects, so at least some scientific advice was potentially available to President Reagan. Several lobbying groups, such as High Frontier, were promoting particular space-weapon schemes, and were therefore anxious to proffer 'expert' advice. It seems likely, however, that myths about the capacity of science were as important, if not more so, as any proven scientific progress in this area. Faith in the prowess of science — particularly American science — is a political force in the United States, as it is in the Soviet Union. Myths about science, combined with a cultural predisposition to look for technical fixes, were probably just as important to the SDI programme as any kind of expert advice. This may well be true of many other defence policies and programmes.

Yet, powerful as the American President is, the SDI programme must have been sustained by more than the 'great communicator's' faith in the potency of science. Scientists working outside military institutions (and many insiders also) have been almost universally sceptical about the feasibility of establishing a leak-proof shield. Given that this is the case, how can one explain the continuation of the programme and the wide support for it in the American bureaucracies

Introduction

and amongst the general public? Each of the available explanations locates the role of science and expert advice differently.

A number of explanations do indeed see science as the servant of policy in this case; but the relevant policy is a hidden or unstated rather than an overt one. Perhaps there is a covert policy of developing an American first strike capacity, with ballistic missile defences designed to intercept Soviet retaliatory attacks using any remaining missiles that escaped pre-emptive destruction by the United States. Perhaps SDI is really a programme designed to develop space weapons for attacking Soviet satellites or ground targets, or laser weapons for the conventional battlefield. Maybe it is part of a clever political strategy to legitimise the breaking of the anti-ballistic missile (ABM) treaty or for securing federal funding for commercially important technologies. Or possibly the initiative was intended to harness popular faith in American science to lend credibility to a political initiative aimed at outflanking the national nuclear freeze movement.

Each of these interpretations may have some explanatory value. However, other interpretations have identified the role of science and scientific institutions in the SDI differently. They note that the budgets for most of the technologies funded through the SDI programme were already expanding rapidly before March 1983 through the patronage of the Defense Advanced Research Projects Agency (DARPA) and the interest of the large scientific and military bureaucracies. In this view, the constant, and most significant factor, is not so much official policy-making as the continuing and semi-independent development of the technologies.

From this perspective the development of military science and technology is not the servant of security policy. Rather it is a relatively autonomous process which generates its own powerful pressures for funding and for incorporation into military procurement programmes. Military and security policy are the manifestations of ideology or *post hoc* rationalisation. Supply models of this kind emphasise factors such as the technological imperative, the power of the scientific bureaucracies within the decision-making process, the role of the military–industrial complex and the follow-on imperative.

Even in its own terms, the concept of science as the servant of a specified military policy cannot be satisfactory. Scientific work in unrelated civil and military areas is bound to structure and shape the development of mission-oriented research. Furthermore, the results of development work on one system will often reveal possibilities for new weapons systems for distinct military roles. More often than not decisions about whether or not to pursue these possibilities will be taken within the scientific or the military bureaucracies themselves. It is not until major increases in funding are required that senior political or military authorities need to be involved, and they will tend to take the advice of subordinates. By that stage lobbies are likely to have developed around the new system, according to its commercial potential and its perceived implications for the prestige and funding of the various branches of the military. Since Western military policy in particular relies so much on maintaining a technological edge over the Soviet Union, there is little prospect of the necessary restrictions on R & D funding being implemented to subordinate military R & D to formulated policy.

It would be wrong, however, to exaggerate the relative autonomy or power of the technological imperative. Technologies are developed and shaped within a social context. Fallows, for example, has described how the designs of the M-16 rifle and the F-16 fighter aircraft were so perverted by bureaucratic pressures that promising technological successes were neutralised.[3]

In the case of the M-16 rifle, a new design that promised to be more lethal and better adapted to wartime requirements than existing United States Army models was reduced by the Army Ordnance Corps to an unreliable and maladapted weapon. This degradation was, apparently, brought about partly by resentment that an outside design was favoured over in-house models. It was also caused by adherence to established scientific criteria for rifle design which were bolstered by the bureaucratic ethos of designing rifles for long-range target shooting rather than for wartime conditions. These 'scientific' criteria were used to justify the amending of the new design to make it approximate to existing rifle designs, thereby destroying the design coherence of the weapon and producing an unreliable weapon that was to become notorious during the Vietnam War for jamming in combat.

The development of the F-16 revealed similar processes at work. A new design for a relatively cheap fighter aircraft was developed by a team critical of the Air Force's traditional design philosophy. The design emphasised capabilities that operational analyses showed to be critical to a fighter's combat effectiveness and jettisoned some of the normal specifications demanded, such as high top speed, which were shown to be less critical. Largely through the patronage of Defense Secretary Schlesinger's office, the Air Force was forced to accept this design. However, in the final stages of development, Air Force design teams gained control of the project and re-incorporated many of the features traditionally favoured by the Air Force, thus robbing the design of many of its advantages.

Many other examples of the bureaucratic and social shaping of military technologies exist. The way in which new military technologies are integrated into defence preparations, as well as the technological developments themselves, are strongly socially conditioned. For example, the role of the cavalry in European armies actually increased in the decades before the First World War, although the development of the machine gun heightened the vulnerability of such forces. All decisions about policies and technologies are taken in the context of conflicting bureaucratic, commercial and political interests, competing myths and values and established 'scientific' or 'expert' assumptions.

In defence, as in other areas of public policy, expert analysis and advice enters the political arena as one element in a highly complex process. The greater the claim of an analysis to scientific status, the more politically prized it is. Because science has acquired such prestige, with a reputation for producing reliable and effective knowledge and for being neutral and value-free, defence-relevant science or supposedly scientific analyses are powerful weapons in the political process.

First, science is used to show that the position being argued is based upon valid methods—that is, on methods that the relevant scientific sub-community would endorse, while refusing similar endorsement to other viewpoints. Secondly, the knowledge is appropriated by another social group and used by them as part of a campaign to discredit opponents in the larger social or political process, while, at the same time, promoting their own. When science is used in this way, much of the nuance with which scientists qualify their arguments is lost and, as a result, scientific data often appear to be being used as an intellectual blunderbuss with which to destroy an opponent rather than as an element in a rational debate.[4]

In the debates surrounding the SDI programme, scientific analyses have played a prominent role. Scientists critical of the programme have analysed various ballistic defence options and calculated that the defences would be extremely expensive, infeasible, unreliable or destabilising. Organisations of scientists such as the Union of Concerned Scientists, the Federation of American Scientists, Pugwash and Scientists against Nuclear Arms have all participated. The analyses they have produced have been prominent in public and Congressional debates.

In turn, pro-SDI scientists (mainly from within the United States defence establishment) have produced counter analyses. The Pentagon has conducted a number of highly publicised tests, which purportedly demonstrate the feasibility of some of the relevant technologies. These debates illustrate features common to many public scientific debates relevant to defence. There is rarely any serious attempt for scientists holding different positions to engage with each other. Whatever the scientists intended, scientific analysis is mainly used to bolster one or other political position. The influence of one scientific analysis compared with another often has more to do with the political importance of the groups that adopt it than with its intrinsic power or validity. The politicians who are so ready to accord prestige to science when it legitimises their position are quick to withdraw it if the analyses are politically embarrassing.

The simplest strategy for handling uncomfortable analyses is to ignore them. Established knowledge or widely accepted interpretations can coexist quite happily with contradictory evidence. It is only when the contradictions are forced to the attention of decision makers or the general public that the established mythologies become vulnerable. The government and other powerful interest groups can often afford to restate arguments or theories long after they have been discredited academically, because they still sound plausible to the non-expert.

Even if the critiques and alternative analyses are heard, most people (be they the general public, political leaders or officials) have neither the time nor the expertise to review the evidence and reach an independent assessment. Mostly it is a question of deciding who to believe. Thus scientific prestige, expert reports and such like are very important. Furthermore, the options are often considered in detail in closed committees to which critics and outside experts have little or no access.

The defence establishment usually has an important advantage over civilian critics in debates relevant to national security interests. It can claim privileged knowledge based on access to classified information. Thus scientists sceptical about the feasibility of SDI have repeatedly been assured that their analyses have been based on out-of-date data and that recently classified experiments showed that their conclusions were wrong. Frequently the scientific critiques of SDI were very robust in their assumptions and unlikely to be invalidated by classified work. However, the appeal to privileged information is hard to gainsay.

Uncomfortable scientific analyses can also be discredited or ignored on the grounds of uncertainty. All analyses involve uncertainties and most also have debatable implications. This means that the analysis is open to legitimate criticism and that attention may be focused on the weaknesses of unwelcome scientific analyses, even if the established mythology rests on even flimsier foundations.

Knowledege about the likely effects of using nuclear weapons should be an essential factor to be considered in nuclear policy-making. Scientists have calculated during the 1980s that a major nuclear war could have catastrophic consequences on the Earth's climate. The possibility of 'nuclear winter' suggests that the possession of nuclear weapons poses even greater risks than was previously thought, affecting non-combatant as well as combatant nations. Many in the scientific community therefore believed that the possibility of nuclear winter should provoke a major re-assessment of nuclear arms and disarmament policy. Chapter 1 examines how the nuclear winter hypothesis was received and used by scientists, interest groups, defence establishments and governments. The impact of the hypothesis on public attitudes to nuclear questions assured that the potential implications of the hypothesis would be hotly debated. But it becomes clear that the way the debates have been conducted has been shaped, if not dominated, by the existing interests groups and institutional preferences. Scientists are not entirely politically naive, however, and the chapter also looks at how they presented their work in order to affect the way it was used in policy debates.

Chapter 2 is concerned with one aspect of the connection between myths, science and the SDI programme, a relationship which, as we have indicated, involves many issues. David Carlton focuses on the origins of President Reagan's policy initiative announced in his speech on 23 March 1983. What role did scientific and technological advice play in the formulation of Reagan's vision of establishing a leak-proof shield against missile attack? What influence did the several lobbying groups promoting ballistic missile defences have on this vision?

Carlton concludes that Reagan received no scientific nor technological advice encouraging him to hold out the hope of a 'leak-proof' shield. Indeed, no significant number of independent scientists, whether for or against deploying limited strategic defences, have dissented from the judgement of Richard DeLauer, former Under Secretary of Defense for Engineering, that 'There's no way an enemy can't overwhelm your defences if he wants to badly enough.'[5] That does not mean, of course, that the programme initiated after that speech will not profoundly affect the future development of military technology. The reason why the Soviet Union so much fears SDI may be, not that they fear that leak-proof shields may be feasible, but that even the present reduced level of SDI spending on military research programmes is likely to produce, not only better limited defences, but other military innovations as yet unforeseen. As the United States Army learnt long ago, scientific research is an unpredictable business.

It is precisely this aspect of scientific research—its tendency to throw up technological innovations not specified in advance by the defined needs of strategy—and the possible subsequent commitment of their creators to see those innovations developed and deployed which is the subject of the chapter by Margaret Blunden, Owen Greene and John Naughton, which is an analysis of the Zuckerman hypothesis. Lord Zuckerman, former Chief Scientific Adviser to the British Government, contributed colourfully to one variant of the bureaucratic model of the arms race, conceived as being fuelled by domestically based interests pressurising their government to arm. The various pressure groups which have been highlighted in this context include the military, civil servants and executives in the defence industries. Zuckerman focuses rather on the characteristic role of the professional scientific community, working in teams on defence-related research in laboratories. He argues that 'the arms race is undoubtedly fuelled by the technicians in government laboratories and in the industries which produce the armaments.'[5]

Many people have been convinced by the Zuckerman hypothesis. It is now established as a powerful explanation of the dynamics of the arms race in the mythology of the general public and many defence analysts alike. In this chapter the Zuckerman hypothesis is first placed in the context of other explanations of these dynamics, and then tested out in the case of some significant steps in the arms race, such as the hydrogen bomb, MIRV technology, and the MX and the cruise missile in the United States, and the Chevaline programme in the United Kingdom. If the hypothesis applies anywhere, one could expect it to apply in cases such as these, yet it becomes clear that the reality is much more complex. To focus on science and scientific bureaucracies may be seriously misleading.

Of course, the secrecy involved in military R & D and weapons procurement makes investigations into such issues difficult and frequently inconclusive. In this context the theories of scientists such as Zuckerman and Herbert York, who have held senior positions within the defence establishment and who have had wide access to classified material, command attention because of their privileged knowledge. However, secrecy and privileged knowledge are themselves important ingredients in the way policy debates about defence issues are conducted. They may also go some way to explaining the popular appeal of certain mythologies about the role of science in the arms race.

The relationship between secrecy, privileged knowledge, authority and access on the one hand, and the making of policy on the other, is taken up in a more sustained way in the second part of this book, which widens the canvass beyond defence policy alone to related civil and military decision making and beyond physical science to both the physical and the social sciences, as possible inputs to the policy-making process. Whereas much of Part I focused mostly on the United States, Part II is more concerned with the United Kingdom.

Chapter 4 looks at aspects of the relationship between the British civil nuclear programme and British nuclear weapons development, with its historical dependence on an exceptionally close relationship with the United States. In this case, as in others we have looked at, the relationship between scientifically established knowledge and the making of policy, military and civil, has been a complex, if not tenuous one, complicated *inter alia* by geo-strategic factors, ideology and bureaucratic momentum. Establishing exactly what policies have been pursued in this area, and what effects the military links have had on the

Introduction

overall management of the civil nuclear power programme, is itself a complicated question. This chapter traces the source and perpetuation of myths in this field, both those deliberately created and those, more subtle and pervasive, which have been allowed to develop of their own accord, born of uncorrected public misunderstandings and the continuing obscurity of vital information, only technically in the public domain.

The complex interaction between civil and military policy making, where a common technology underlies both areas, is as apparent in electronics, the subject of the next chapter, as it is in nuclear power. As far as military electronics is concerned, the pace of technological change has strategic and budgetary implications of such scale and complexity that Western policymakers have yet to come to terms with them. The effect of military R & D on the health of the civil electronics sector and how best to reconcile civil and military interests in this sector, which is crucial to both of them, are contentious questions. Here it is, of course, not the physical sciences but the less prestigious social sciences which might be expected to have a contribution to make to policy. As we have seen, the way in which research in the physical sciences, with their reputation for being neutral and value-free, enters the political arena is far from straightforward, constituting but one input into a highly complex process. As this electronics case study demonstrates, the relationship between social science, political ideology and government policies is still more problematic. A certain amount of research has been done from a variety of perspectives in Britain and elsewhere, and associated theories formulated, on the relationship between civil and military R & D and its effect on the economy as a whole. This chapter explores the often tangential relationship between the several studies and reports prepared and the research undertaken and the policies actually pursued. The mechanisms by which policy makers may see and respond to reports, especially those with a technical dimension, are not as well developed in the United Kingdom as in some other Western countries. The policies pursued, which themselves constitute important real time experiments, are not often monitored with sufficient detail and accuracy to further knowledge. Such knowledge might, after all, turn out to be politically unwelcome.

Clearly defence policy making in Britain does have its own knowledge base. What this has traditionally derived from, what its limitations are and how it is changing is the subject of the last chapter. The British constitution and traditional practice have confined the boundaries of influence over defence decision making very narrowly in the past, creating a system within which few were in a position to challenge the expertise of defence civil servants and senior military personnel. The idea that ministers, with their own political agendas, have sufficient knowledge, managerial skill or political will to control the activities of their departments, has been widely and popularly debunked in Britain. At the Ministry of Defence in particular the lengthening time scales of weapons procurement, the increasing technical complexity of weapons development and the changing environment of military policies have made control by ministers (who generally have no special expertise or even interest in defence, lack managerial experience and often serve short terms of office) particularly difficult. Parliament, for its part, has been able, through the work of the House of Commons Defence Committee, first formed in 1979, to extend its knowledge base, although it still has, unlike the French, no equivalent of the American Office of Technology Assessment to help it to disentangle difficult technical issues. As the defence consensus has crumbled in Britain, the Ministry of Defence, for its part, has seen advantages in trying to forge closer links with universities and defence institutes. The possible effect of these on policy or on research has yet to become apparent.

Each of the chapters in this book deals with important issues and contains original material which we expect to be of interest to specialists and non-specialists alike. Each, in its different way, explores the relationships between science, myths, expert knowledge and the policy-making process. The book aims to illuminate a number of controversial issues and to arrive at a fuller appreciation of the complexity of such relationships.

Notes

1. Don K. Price, 'The scientific establishment', in Robert Gilpin and Christopher Wright (eds.), *Scientists and National Policy-Making*, Columbia University Press, 1964, p. 20.
2. Office of Technology Assessment; *Strategic Defenses: Ballistic Missile Defense Technologies; Anti-Satellite Weapons, Countermeasures and Arms Control*, Princeton University Press, 1986, Part 1, p. 298.
3. James Fallows, *National Defense*, Vintage Books, New York, 1981, chapter 4.
4. Michael Gibbons and Philip Gummett, *Science, Technology and Society Today*, Manchester University Press, 1984, pp. 10–1.
5. Solly Zuckerman, *Nuclear Illusion and Reality*, Collins, London, 1982, p. 103.

1

The Risk of Nuclear Winter: Scientific Research and Policy Debate

OWEN GREENE and JOYCE TAIT

Introduction

Ever since thermonuclear warheads and long-range missiles were developed, it has been widely accepted that nuclear war would be an unprecedented catastrophe for every country involved. Beyond this, however, questions relating to the consequences of nuclear war have been highly controversial.

This controversy is not solely due to scientific uncertainty. Assessments of the likely destructiveness of nuclear war intimately affect attitudes towards nuclear defence policies. Defence policies are designed to reduce the likelihood of enemy attack but there is always the possibility that war might still occur. Choices about defence policy therefore involves a balancing of risks. One policy may provide a strong deterrent, but also make it likely that if war did come it would be enormously destructive. Another policy may imply relatively low levels of destruction in the event of war, but at the cost of increasing the possibility of enemy attack. A policy's risk therefore depends both on its chance of failing and on the potential cost if it did.

Since the consequences of any nuclear war would be appalling, nuclear deterrence policies seem to many people to be highly risky, even if the possibility of such a war actually occurring seems remote. It is in this context that debates about the likely consequences of nuclear war have proved to be controversial. There is the possibility that, if it could be shown that nuclear war would be more destructive than it was previously thought, large numbers of people would change their minds about the acceptability of the risks inherent in nuclear deterrence. Similarly, if people came to believe that nuclear war would not be all that destructive, their confidence in nuclear deterrence might increase.

Some people, believing that a nuclear war would destroy all that they valued, have argued on both moral and pragmatic grounds that it should be avoided at all costs. Since the possibility of unintended or accidental nuclear war cannot be excluded, this implies a rejection of nuclear deterrence policies and a conviction that there are no circumstances in which nuclear weapons should be used. For these 'nuclear pacifists', the possibility of 'nuclear winter'* posed few challenges to their established beliefs. For them, it drew attention to yet another devastating consequence of nuclear war and reinforced their support for complete nuclear disarmament.

At the other extreme, there are those in the West who clearly believe that it would be 'better to be dead than red' and that any weakening of deterrence would greatly increase the likelihood of Soviet attack. Doubtless there are some in the Soviet bloc who hold similar views about capitalism and the West. Such people have staked everything on deterrence, and it is unlikely that any reassessments of the potential consequences of nuclear war would significantly affect their support for counterforce nuclear policies.

However, most people hold intermediate positions, and it is for these that the newly discovered threat of nuclear winter could have the greatest implications. The increased evidence that nuclear war would be unsurvivable for NATO or Warsaw Pact societies could persuade many of their inhabitants that the risks inherent in present nuclear policies are unacceptably great. Support could grow for alternative policies.

Similarly, countries outside the nuclear alliances may previously have regarded the nuclear arms policies of NATO and the Warsaw Pact (and China) as regrettable but of little direct concern to them. But a nuclear winter, threatening combatants and non-combatants alike, could stimulate non-nuclear or non-aligned countries into taking a more active interest in influencing nuclear policies. Finally, there may be people in the United States, Europe or the Soviet Union who were willing to risk the destruction of their own societies, but who believe that it would be immoral, in the uncertain interests of their own society, to threaten the survival of all or most of the global human population, and billions of animals and plants as well.

---

*We shall use 'nuclear winter' as a shorthand term throughout this chapter. However, the reader should note that it refers to a range of different potential effects of nuclear war: major temperature reductions (i.e. more than about 5°C average), reductions in light levels, intermediate and long-term radioactive fallout, global chemical pollution and its impact on the ozone layer, for example. It is not meant to refer only to average temperature reductions of over 15°C. Indeed, as will become clear, most calculations indicate that a major nuclear war could result in inland temperature reductions in the northern hemisphere in the range of 5–30°C.

Thus research on the nuclear winter could, logically, have major policy implications for many people. Not only could these people be expected to take a serious interest in the scientific assessments and debates about policy implications: committed critics or supporters of present nuclear policies could also be expected so to do. This is not because they would expect their own views to change, but because of the potential political impact of the research on those in the centre ground. A committed group would naturally aim to promote scientific research or arguments that seemed to support their case and to play down or discredit those that undermined their position.

The widespread belief that many voters do not seriously attempt to assess risks or balance options, but are liable to switch their support according to which problems are uppermost in their minds, only increases the intensity of debates such as this. Most voters support nuclear disarmament in principle but fear at the same time that any weakness in their country's military forces could result in war or subordination to another state. It is thought that a significant proportion of these people could support either the anti-nuclear movement or existing nuclear deterrence policies, according to whether the horrors of a nuclear war or those of a foreign threat happen to be most vividly in their minds at the time.

Experience seems to indicate that most voters are usually more susceptible to fears of military weakness or impending attack than to images of nuclear war. Attempts to give the risks of nuclear war some psychological reality usually founder. Beyond visions of a blinding white flash or a mushroom cloud, it is hard to do more than conjure up film images or memories of the Second World War; memories, for many in Britain and America, of comradeship and full employment as much as of destruction. To the extent that the risks of nuclear deterrence seem unreal or remain in the background, the general public is unlikely to force a thorough re-examination of nuclear weapons policy.

Images that make the horrors of a nuclear aftermath more vivid and comprehensible might therefore be expected to have a major impact on public opinion. Controversial drama documentaries about the potential effects of nuclear war, such as BBC Television's 'Threads', derived much of their impact from their portrayal of comprehensible horrors, such as the threat of a total breakdown in law and order, or the misery of dying hungry or in pain in filthy conditions. The possibility of a climatic catastrophe has a similar vividness. People may feel that they can begin to comprehend the impact of cold and dark on any initial survivors and on agriculture, animals and plants. A climatic catastrophe could also affect people and nations who might hope to escape the direct effects of nuclear attack. The prospect of nuclear winter undermines civil defence programmes and the widespread 'survivalist' or 'back-to-nature' fantasies that many have found comforting.

Policy makers and protagonists in the nuclear debate were quick to appreciate the potential impact of the nuclear winter hypothesis on public opinion. From the moment that the first scientific research results were published in 1983 they were the subject of intense political and scientific controversy. This was partly because of their potential effect on the public acceptance of the risks of nuclear deterrence, and partly because the possibility of climatic catastrophe could force a re-examination of many aspects of established targeting strategy and nuclear weapons programmes.

Thus the first reaction of many commentators and experts was to cast doubt on the significance of the research and to attempt to dismiss it as bad or politically-motivated science. However, it was not as easily dismissed as had been previous predictions of long-term global collapse that rested on assumptions about complex, synergistic processes involving social and ecological sciences. In contrast, the possibility that nuclear war could trigger major changes in the climate arose from work in the more prestigious quantitative disciplines, with their presumption of scientific rigour. For the first time such research indicated a mechanism by which *billions* of people could die as a result of nuclear war.

Policy debates about the risks presented by hazardous technological developments are becoming increasingly sophisticated and research on risk assessment has made us more aware of sources of controversy and disagreement. We react differently , for example, to risks undertaken voluntarily, for our own benefit, compared to those imposed upon us for someone else's benefit; to insidious, chronic hazards, compared to those that are obvious and instant in their effects; and to events that kill large numbers of people at one time, compared to those that kill large numbers over a period of time.[1]

Nuclear war is an extreme example of those technological hazards, with a very low probability of occurrence and a large, potentially catastrophic impact. A somewhat less extreme example is the failure of nuclear power or complex chemical plants—both areas in which 'unthinkable' events have actually occurred, as at Chernobyl and Bhopal respectively. These accidents affirmed the fallibility of experts in the civil sphere and must further undermine confidence in the ability of military decision makers to avoid the accidental onset of war.[2] Such worries will be reinforced by the remedies being suggested to avoid further castastrophies of this nature, namely effective monitoring and greater openness to inspection, both of which are still relatively unacceptable in a military context.

As noted above, potentially catastrophic hazards of low probability, such as those associated with nuclear power generation are often less publicly acceptable than risks that lead to significant number of casualties every year. However, wartime conditions produce a mental adjustment of risk acceptability. People who, in peacetime, may refuse to take the risk of being killed while employed as a coal miner (average of 210 deaths per million participant years) or while rock climbing (40 per million participant hours),[3] have readily resigned themselves to the much greater risk of death at the hands of the enemy. In decision making about war, even when it takes place in peacetime, the state has imposed, or planned to impose, extreme risks on the population, with the justification that they are needed for the defence of the nation and its culture. Nuclear war, particularly if accompanied by nuclear winter, would remove this justification. It would destroy the very society we are trying to defend. Hence the nervousness of policy makers about the impact on public opinion of predictions that alert people to the destructiveness of nuclear war.

In the early 1980s the probability of nuclear war was, for a variety of reasons, perceived to be increasing. Simultaneously, the nuclear winter hypothesis suggested that its consequences would be worse than previously imagined. This could and should have provoked a serious reassessment of nuclear policy in relation to such risks. Ideally, this process would involve quantitative risk assessment which stresses the importance of scientific rigour[4] and a clear separation of facts and values.[5] In practice, however, decisions need to be made in the context of inadequate information. Frequently scientific understanding among the public and policy makers is poor; questions of risk and uncertainty, fact and value become confused; and political demands for inappropriate scientific precision may distort the presentation of scientific advice.[6]

This chapter outlines the development of research on the potential climatic effects of nuclear war and examines how this scientific work has been shaped by and used in policy debates and policy making. An outline of the development of the nuclear winter hypothesis is followed by a discussion of its potential policy implications and of its actual impact on policy and policy debates in the United States, the United Kingdom, the Soviet Union and elsewhere. The chapter does not present a case for one particular set of policy implications. Rather it describes how the research and its potential implications were discussed and perceived by the main participants in the debate. The story, interesting in its own right, highlights many aspects of the ways scientific research may affect policy debates, and expert knowledge and myths contribute to policy making.

Scientific research on climatic consequences of nuclear war

Beginnings

A nuclear exchange involving about one third of the world's nuclear arsenal (5,000–6,000 megatons) would eject about one billion tonnes of dust high into the atmosphere and could ignite fires over an area of more than one million square kilometres, of which perhaps 250,000 square kilometres would be in urban areas. These fires would generate vast quantities of smoke, of which it is estimated that between 50 and 150 million tonnes could rise into the upper atmosphere. This smoke and dust would be spread by the wind. Within one to three weeks the sky over much of the northern hemisphere (and perhaps some of the southern hemisphere) could be covered by a cloud of dust, ash and black soot. If so, energy from the sun would be absorbed or scattered in the upper atmosphere, reducing the light and heat received at the Earth's surface and lower atmosphere.[7]

Many of the processes involved in generating and spreading such a cloud are complex and poorly understood, and difficult scientific problems have to be tackled in any comprehensive analysis of the likely characteristics of the cloud and of its possible impact on the climate. Large numbers of scientists have been working on these problems and the research programmes continue. However, it is evident from our knowledge about the sheer scale of the potential disruption to the absorption of light and heat from the sun that there is major cause for concern.

Surprisingly, the first rough calculations analysing these effects were not carried out until 1982, although it was well-known, since the atomic attack on Hiroshima, that heat from nuclear fireballs could ignite fires over vast areas. Scientists studying the potential consequences of nuclear war had neglected smoke—perhaps the key factor—for over 30 years.

In 1980 the editors of *Ambio*, the environmental journal of the Royal Swedish Academy of Sciences, had commissioned a special double issue on the possible human and ecological consequences of nuclear war. They brought in five special advisers to help to structure the issue and to commission the articles, including Professor Joseph Rotblat, ex-General Secretary of Pugwash, and Frank Barnaby, ex-Director of the Stockholm International Peace Research Institute. Both are acknowledged experts on nuclear issues, but are also well-known for activities, stretching back to the 1950s, aimed at controlling and reversing the nuclear arms race.

A Dutch scientist, Paul Crutzen, and an American, John Birks, were commissioned to investigate the likely effects of nuclear war on the atmosphere. It was envisaged that they would concentrate on the possible effect on the ozone layer, as many others had done before. Crutzen had been one of the first scientists to draw attention to this danger over a decade earlier. However, it occurred to the two scientists that smoke from fires ignited by the heat of nuclear fireballs could affect the chemical processes that they were studying. Rough calculations showed that there could be enough smoke to blot out nearly all of the sunlight from half the earth for weeks on end.[8] They did not have time to explore properly the implications of this discovery in their *Ambio* article[9] but they did draw attention to the possible dramatic reductions in sunlight and the dangers of poisonous chemicals in the smoke.

The next significant development took place at a conference organised by the International Physicians for the Prevention of Nuclear War (IPPNW) in April 1982, when Rotblat referred to Crutzen and Birk's work. Professor Carl Sagan, an American planetary physicist already well known as a populariser of science, was in the audience.[10] His interest in the effects of dust storms on Mars alerted him to the potential impact of clouds of smoke and dust on temperatures at the Earth's surface. After a brief discussion with Rotblat, Sagan read Crutzen and Birk's work before returning to the United States. There he assembled a team of scientists to investigate the problems further. It included Richard Turco, who was already interested in the issue as a result of reading a preprint of the *Ambio* article.

This team produced the first detailed quantitative analysis of the potential global effects of smoke and dust clouds after nuclear war. It also coined the term 'nuclear winter' and presented the hypothesis to the scientific community and the general public. It brought to the subject both a range of expertise and a set of computer models developed to study the effects of dust and pollutants in the atmosphere of the Earth and other planets. Their names were Richard Turco, Owen Toon, Thomas Ackerman, James Pollack and Carl Sagan—TTAPS for short.

By early 1983 TTAPS had carried out their first full calculations. These indicated that inland surface temperatures in the northern hemisphere could fall by 20–40°C within a few days of a major nuclear war, and take up to one year to return to within a few degrees of normal temperatures. The amount of sunlight reaching the ground could be reduced to a few per cent of normal and even at noon many places could be as dark as on a moonlit night.[11]

Launching the nuclear winter hypothesis

The TTAPS group understood that their calculations were potentially highly significant, and were bound to provoke controversy. The nuclear freeze movement was at that time approaching the peak of its influence in the United States, as were the anti-nuclear movements in Europe. Scientific caution and political acumen led them to arrange for their results to be thoroughly reviewed by dozens of physical scientists during the spring and summer of 1983. TTAPS refined and revised their work in the light of these comments.

Sagan, in particular, began to prepare a 'package' so that the nuclear winter hypothesis would have maximum impact when it was launched. For example, in April 1983, the TTAPS results were presented to a large group of distinguished biologists and ecologists at a meeting in Cambridge, Massachusetts. This group prepared an ecological analysis of how the climatic changes could affect plants and animals. This was then circulated for review. A slide show and a dramatic five-minute video programme were prepared for nationwide and international distribution.

The nuclear winter research was first publicly presented at the 'Conference on the World after Nuclear War', sponsored by more than 30 scientific and environmental bodies, held in Washington, DC on 31 October and 1 November 1983. Some 500 scientists, journalists and others from 17 countries attended. Senior scientists from the United States National Academy of Sciences and other scientific academies were integrally involved. Critics of the nuclear winter hypothesis could be forgiven for thinking that they were even then opposed by the bulk of the scientific community.

Significantly, Sagan had taken steps to involve Soviet scientists at the earliest possible stage. Early in 1983 he contacted a number of influential Russians, including Academician E. P. Velikhov (Vice President of the Soviet Academy of Sciences, chairman of the 'Soviet Scientists' Committee for the Defence of Peace against Nuclear Threat and a key scientific adviser to the Soviet leadership) and Dr E. I. Chazov (then personal physician to the General Secretary of the Soviet Communist Party and joint-President of IPPNW). By the time of the Washington conference, some Soviet climatologists had qualitatively reproduced some of the key TTAPS results, using much cruder assumptions and computer models, and the nuclear winter phenomenon had received the stamp of approval of the Soviet Academy of Sciences. On the second day of the conference, a teleconference was organised between the participants in Washington and Soviet academicians in Moscow. This enhanced the impression of an international scientific consensus. The whole international scientific establishment seemed to be joining together to warn humanity about the newly discovered risk of nuclear winter.

The launch of the nuclear winter hypothesis was therefore impressive. Care had been given to ensuring that the research could not be easily dismissed and that a major political impact was achieved. It is worth noting that, although TTAPS's nuclear winter research had been widely reviewed within an informal scientific network by October 1983, no papers on their calculations had yet been published. TTAPS published a short paper in *Nature* in November accompanied by a study of the potential biological consequences of their results carried out by the biologists referred to above.[12]

The scientists' confidence in the TTAPS group and the robust nature of its conclusions appears to have been justified. Four years of intensive scientific work has so far tended to confirm TTAPS's qualitative results. However, in November 1983 the nuclear winter hypothesis rested largely on Crutzen and Birk's earlier paper and TTAPS's work. Some researchers at the National Center for Atmospheric Research in Colorado and the Lawrence Livermore Laboratory were beginning to generate similar results using two- and three-dimensional computer models of the Earth's atmosphere (i.e. different from TTAPS's approach).[13] However, the independently produced research corroborating TTAPS was new and unpublished. Although the Soviet results were received politely at the conference and were important for gaining the early support of the Soviet Academy of Sciences, the modelling was crude and did not justify increased confidence in the scientific validity of the results.

The Washington conference organisers, aware of the danger of the research being attacked as politically motivated or straying from areas of scientific competence, avoided discussions about policy implications. Their report states that:

> The Steering Committee, in preparing for the scientific conference programme, decided to consider only the physical, atmospheric and biological consequences of nuclear war ... the inclusion of other considerations such as nuclear strategy and economic, social and political implications would detract from the central scientific message.[14]

However, awareness of the potential policy implications and the political context penetrated every aspect of the early development of nuclear winter research. From the beginning particular efforts were made to involve Soviet scientists and to emphasise the advantages of collaboration and the global nature of the problems. The special issue of *Ambio* was stimulated by increased public concern about nuclear war. In their introduction to the issue, the editors finished with a quotation from the 1980 UN Report on Nuclear Weapons: 'As there is no guarantee that the risk of war can be avoided, the need for nuclear disarmament is imperative.'[15] Key individuals had met at IPPNW meetings and had longstanding concern about the nuclear arms race. Carl Sagan and Paul Ehrlich (who headed the group of biologists) and several of the other scientists involved had long experience of participating in policy debates, and Velikhov and Chazov in the Soviet Union were closely concerned with Soviet politics.

Although detailed discussion of policy implications was avoided at the Washington conference, Sagan had already thought seriously about such issues and about the direction in which he preferred the debates on implications to go. Discussions about policy were therefore included in his overall package. He had prepared an article on the policy implications of nuclear winter which was published in the winter 1983–84 issue of *Foreign Affairs*,[16] produced shorter newspaper and magazine articles and made television appearances.

Furthermore, notions about the best way to assess the policy implications seem to have been central to the scientific approach taken by Ehrlich and his fellow biologists.[17] Whereas TTAPS examined a range of nuclear attack scenarios (from 100 megatons to 10,000 megatons) and several different modelling assumptions, the paper by Ehrlich *et al.* examined only the potential consequences of the worst plausible case—involving maximum temperature drops of 50–60°—and sub-zero temperatures for almost one year. The rationale for this choice was that it was important to consider the worst that could plausibly occur, in order to understand the policy implications. In the face of uncertainties, this would clarify the risks involved in nuclear deterrence policies.[18]

This choice certainly helped to ensure that there was little disagreement with the biologists' overall assessment of the biological and ecological consequences of such a climatic disturbance. As Ehrlich has said, it was like assembling a group of neurosurgeons to assess what would happen if a patient put a 12-bore shotgun in his mouth and pulled the trigger—they may disagree about the details, but the overall consequences for the patient would be clearly terminal. The problem was, however, that many critics could (and did) disagree with Ehrlich's approach to the policy implications and insisted that other, more probable scenarios were more relevant. In the absence of serious assessments of the biological and ecological consequences of milder nuclear winter scenarios, such critics were free to claim that these would be much more survivable.

The scientific status of the nuclear winter hypothesis

Since November 1983 there has been intense scientific interest in research on topics relevant to nuclear winter. Hundreds of scientists have written scientific papers and dozens of scientific conferences and meetings have been organised each year throughout the world. Interest in the issue has encouraged many of these scientists to participate in public debates and appear in the media. Critics of the nuclear winter hypothesis have publicised any research that appeared to indicate that the climatic disturbances could be less than TTAPS indicated. As journalists searched for a new angle to their new stories, misleading reports abounded that the possibility of a climatic catastrophe had either been disproved or re-established.[19] Some scientists, encouraged by the idea that their research or comments could be of significance to policy makers (and, perhaps, bring fleeting fame), unwisely rushed into print with poorly substantiated views.[20] Many people who had some claim to technical competence but did not keep up with the scientific literature continued to worry at the details of the original TTAPS paper as if it still stood alone as the foundation of nuclear winter research.[21] The nuclear winter issue assumed all the trappings of a scientific bandwagon rolling at full speed.

Scientific research on the subject came under almost permanent review by scientific institutions and academics. Throughout 1984 a Committee of the National Research Council, part of the United States National Academy of Sciences (NAS), examined new scientific evidence on the effects on the atmosphere of a major nuclear exchange and produced a report early in 1985.[22] At about the same time the Royal Society of Canada produced a report on both the potential physical effects and the biological consequences.[23] Perhaps most importantly, in 1983 the International Council of Scientific Unions (which includes all the main national academies of science) set up a steering committee to co-ordinate and review international research on the environmental consequences of nuclear war. This SCOPE (Scientific Committee on Problems of the Environment) sub-committee organised a series of specialist scientific workshops and review meetings throughout the world and encouraged, co-ordinated and occasionally commissioned research. The report produced in the late summer of 1985, the most authoritative and comprehensive review of the subject so far, involved more than 300 scientists from 30 countries.[24]

The reports differ somewhat in their detailed assessments of particular issues, but are in reasonable agreement about the probable major global effects of large-scale nuclear war. There is now little doubt that it would result in vast quantities of smoke and dust being sent high into the atmosphere, or that this would reduce surface temperatures and sunlight over large parts of the Earth's surface. However, there is still uncertainty as to the magnitude, extent and duration of these effects.

Sunlight at ground level could be reduced, on average, to 10 per cent of normal values over the northern mid-latitudes (which stretch from Alaska or Norway in the north to Miami or Morocco in the south). However, these light levels could fluctuate as particularly dense or thin patches of cloud passed overhead. Sunlight would be significantly but less substantially reduced over the rest of the northern hemisphere and the southern tropics.

In summer, temperature reductions of 20–40°C within a few days appear to be quite possible in the continental interiors in the northern mid-latitudes. In coastal areas the average temperature drop would be less severe but temperature fluctuations could be extreme. Within a few weeks smoke particles would be expected to spread throughout the northern hemisphere, quite possibly resulting in fluctuating temperature reductions of as much as 15°C.

Average summertime land temperatures could therefore drop to levels typical of autumn or early winter for weeks or months, with continental interiors being subjected to short periods of 'midwinter' temperatures. Tropical areas could experience periods of cold, perhaps with outbreaks of frost, such as have never happened before. Strands of smoke could be carried into the southern hemisphere within one or two weeks, leading to episodic temperature reductions of up to 10°C. Weather systems would also be seriously disrupted. For example, the summer monsoon rains in Africa and parts of Asia could be eliminated and rainfall patterns could change radically. In winter temperature reductions would be less severe and the southern hemisphere is less likely to be disrupted.

It is not clear for how long these effects would continue. Most studies indicate that acute effects would pass after a few weeks or months, perhaps returning to normal (on average) within a year or two. However, the rate at which smoke particles would be removed from the middle and the upper atmosphere is subject to many uncertainties and most of the studies omit important processes which could greatly affect the conclusion. Computer models of weather systems are not yet sufficiently sophisticated to cope with them. There are reasons to fear that the severe climatic disruption could be long lasting (perhaps one to three years) but other factors could make this disruption more short-lived.

The SCOPE report also estimates that ozone concentrations in the stratosphere could be reduced by 10–30 per cent, leading to increases in the intensity of biologically-active UV-B radiation at the Earth's surface of 30–100 per cent. Large amounts of poisonous chemicals would also be released as a result of fires in urban and industrial areas.

The account above describes what appear to be, on the basis of present scientific knowledge, the most likely consequences of a large-scale nuclear attack involving the detonation of about 5000 megatons or more. However, the SCOPE and other reports stressed the many uncertainties underlying their calculations. It is possible that the likely inland temperature reductions after a large-scale nuclear war could be revised in the near future to an average of 5–10°C, a 'nuclear autumn' rather than a nuclear winter. Equally, further research could show that the catastrophe would be even more severe and long-lasting than present calculations indicate.

It is important to recognise that many uncertainties can never be resolved. The nuclear war scenario, the time of year and other key factors are intrinsically unclear. In the absence of the ultimate experiment, nuclear war itself, there will always be debate about how the sudden introduction of vast quantities of smoke and dust into the atmosphere will affect atmospheric processes and the climate. Even after a nuclear war (to carry scientific inquiry to extreme limits) there would be doubt about the reproducibility of the observed effects in any future war. Uncertainty will therefore remain intrinsic to nuclear winter research.

Many of the debates among physical scientists appeared irrelevant to biologists studying the potential consequences of the climatic changes. The question of whether the nuclear winter may turn out to be a nuclear autumn hardly interests agriculturalists, who recognise that a 2-3°C average temperature reduction during the growing season would be an unprecedented disaster. Volume II of the SCOPE report presented detailed evidence for the view that a nuclear war followed by a 3-5°C average temperature reduction in the northern hemisphere (and the changes in precipitation patterns inevitably associated with this) could destroy agricultural production in the northern hemisphere for one year or longer.[25] Industrial and agricultural production in the southern hemisphere would also be severely disrupted because of the interdependence of the global economy and also, perhaps, because of the effects of smoke and dust carried from northern regions. Since over 88 per cent of the world's population live in the northern hemisphere (with about 42 per cent in the mid-latitude region),[26] it is clear why scientists believe that even a nuclear autumn could lead to the deaths of billions of people. The ecological impact of such climatic changes, combined with effects on the ozone layer, would be disastrous, long-lasting and perhaps irreversible.

Potential policy implications

It has become an accepted part of the political process for scientists, acting as legitimate and authoritative sources of information, to speak out in the public interest.[27] If their claims are taken seriously, a debate begins in which rival experts and interest groups are encouraged to present their views to political arbiters. The tone of the expert debate is still overtly scientific but its real nature has undergone a significant change. The rules of the policy-debating game bear little relationship to those of academic life.[28] For example, pressure groups tend to consider arguments more in terms of whether they support each group's interests than in terms of their scientific merits. There is a continual (and often unscrupulous) struggle to set the agenda of the debate. Frequently experts are personally attacked as protagonists seek to undermine their credibility. Thus it is one thing for scientists to initiate a policy debate and quite another to sustain it or to influence its outcome. In the case of nuclear winter the potential policy implications are debatable and some people have even questioned whether there are any such implications.

The final paragraph of the proceedings of the 1983 Washington Conference on the World after Nuclear War concluded that:

> The scientific part of the process must continue so that the uncertainties are reduced. But we already know enough of the risks to recognize that it is imperative, in the name of humanity, to accelerate the search for world security in the public domain.[29]

However, the prestigious international and North American scientific institutions decided, in the main, that they should stick to the production of technical reports and leave others to interpret them. The scientists involved wanted to ensure that their reports were taken seriously by policy makers and did not want to risk the reports' being dismissed with accusations that scientists were going beyond the boundaries of their competence. Similar concerns led to a concentration on 'hard' natural scientific research, which has greater public prestige than the predictions of anthropologists, social scientists and psychologists.[30]

The publications of the reports by SCOPE, the National Research Council (NRC) and other scientific institutions reinforced the seriousness of the nuclear winter issue and confirmed the possibility of climatic catastrophe. They helped to keep the debate on the political agenda, but they did not themselves contribute to the analysis of the policy implications. The NRC, the SCOPE and the Royal Society of Canada's reports mainly confined themselves to emphasising their collective belief that, whatever the policy implications might be, they were very significant.

The NRC report, produced on behalf of the US National Academy of Science, was particularly cautious, possibly fearing that the political environment in the United States in the mid-1980s might be hostile to any observations about policy. After stating that the possibility of major climatic changes is 'of sufficient national and international concern that a major effort to narrow the scientific uncertainties should be given high priority'[31] it felt that it had to justify producing a report at all: '. . . given the enormous human stakes that may be involved, it may not be advisable to wait until a strong scientific case has been

The Risk of Nuclear Winter   27

assembled before presenting tentative results.' It said of the implications of its technical report that 'Long-term atmospheric consequences imply additional problems that are *not easily mitigated* by prior preparedness and that are *not in harmony with* any notion of rapid postwar restoration of social structure.'[32] (Our emphases.) Such caution seems extreme, bearing in mind the scale of the estimated atmospheric and climatic effects. The NRC did, however, go on to say that:

> [long-term atmospheric consequences] also create an entirely new threat to populations far removed from target areas, and suggest the possibility of additional major risks for any nation that itself initiates use of nuclear weapons, even if nuclear retaliation should somehow be limited.[33]

The SCOPE report, produced on behalf of the International Council of Scientific Unions, went little further in its discussion of the implications of the research. However, the SCOPE committee was more determined to ensure that its report received wide public attention. It took the unusual step of hiring a public relations and lobbying organisation—the Centre for the Consequences of Nuclear War—to publicise the launch of the report in the United States. It also commissioned a writer to produce a more popular paperback version of the report, in the hope of achieving wide sales.[34] The chairman of the SCOPE enquiry, Sir Frederick Warner, FRS, argued in the press release that the report:

> represents the consensus of a prestigious body of scientists. It would be a grave error to ignore their findings. The potential environmental damage of a nuclear war demands that we develop a new perspective when considering such a conflict.[35]

At the press conference itself, Warner added, 'Anybody who thinks they can read this and not draw policy conclusions is making a big mistake.'[36]

In the milder political environment in Canada, the authors of the report for the Royal Society of Canada felt able to express themselves more strongly than the NAS was in the United States. The report begins by taking the position that 'We believe the nuclear winter to be a formidable threat'.[37] It goes on to say that 'the nuclear winter hypothesis does indeed modify the global strategic position'[38] and that strategists had to 'incorporate that judgement into their thinking'.[39]

At a time when Sagan was under attack in the United States, the Canadian report included the following passage:

> Scientists tend to cloak their predictions in a shroud of ifs and buts. On this issue scientists, like Sagan, have spoken out vigorously, and are often accused of exceeding the limits of scientific inference. But the opposite tendency—of ignoring the potential implications of new scientific information—is even more dangerous.[40]

In Canada it was possible for a subcommittee of the Royal Society to make such an observation and still to retain a reputation for scientific authority and neutrality. In Washington, DC, a similar statement by the NAS would probably have been portrayed by many as evidence that its report was politically motivated and unreliable.

On the nuclear winter issue, the scientific academies and institutions have not so much avoided making contentious policy statements as avoided making statements that would be strongly disputed within their own country. In Sweden, New Zealand and the Soviet Union, where, in their different ways, there was little *public* dispute about the importance of achieving nuclear disarmament as a top priority, scientific institutions felt able to go much further in their assertion about the implications of the risk of nuclear winter than similar bodies were in NATO countries.[41]

Protagonists and their arguments

Unlike NAS, SCOPE and other scientific institutions, many individual scientists did become directly involved in policy debates. The NAS itself invited several scientists to present their views about policy implications at the end of its March 1985 conference on nuclear winter.[42] As experts on nuclear winter research, scientists were able to command public and political attention for their views. Dozens of scientists appeared before congressional committees and many more participated in conferences, seminars and television and radio programmes or contributed to newspapers or magazines. They were joined by media commentators, academic strategists, representatives of interest groups, members of the defence community, administration officials, and professional or well-connected lobbyists and self-publicists. Defence think-tanks and consultants, ranging from the prestigious RAND Corporation in California to the so-called 'Belt-way bandits' (the analysts marketing defence advice, of very uneven quality, from their bases on the Washington ring-road) make their living by being alert to potential bandwagons and they quickly recognised the possibilities presented by nuclear winter.

The scientists centrally involved in the 1983 Washington conference had the opportunity to take the initiative in the policy debate, and several of them used it effectively. Carl Sagan's *Foreign Affairs* article, 'Nuclear war and climatic catastrophe: some policy implications'[43] set the agenda for public debate when it was published and it is still influential.

For Sagan and many others, the nuclear winter hypothesis had fundamental policy implications. Whereas previous estimates had indicated that hundreds of millions of people would die as a result of a major nuclear war and that NATO and Warsaw Pact societies would be torn apart, the new research showed that several billions could die and that human civilisation throughout the Earth could be destroyed. By stockpiling thousands of nuclear weapons, the five nuclear weapon states may have inadvertently built a system reminiscent of the doomsday machine—a device designed automatically to destroy most, if not all, human life if ever deterrence fails.[44]

No government, including those of the nuclear weapon states, has ever sanctioned or argued for the construction of such a doomsday machine. The concept plays no role in any of their strategies, and when the possibility was first raised by Herman Kahn in the early 1960s, it was universally rejected. A doomsday machine certainly has some of the characteristics of an 'ideal' deterrent: it is extremely frightening and should ensure that even a fool would be vividly aware of the risk of world destruction and would therefore act cautiously. However, it was agreed that the risks associated with a doomsday machine would be too high, since the possibility of accidents or unintentional war can never be ruled out.

Sagan and others therefore argued that the nuclear weapon states should act quickly to ensure that the climatic doomsday machine was defused, but this need not imply complete nuclear disarmament. A few nuclear explosions would have little or no global effect on climate. Even a war involving several hundred nuclear detonations (killing many millions of people in or near the combatant nations) would be unlikely to affect temperatures in the northern hemisphere by more than an average of 1–2°C. Therefore if the global stockpiles of nuclear weapons could at least be reduced below some 'threshold' level, the risk of climatic catastrophe (involving, for example, average temperature reductions of more than 3–5°C) might be reduced to less unacceptable levels.

In fact the notion of a 'nuclear winter threshold' needs to be used carefully. The scale of nuclear war sufficient to trigger a nuclear winter is uncertain and depends on the season, the targets and many other factors. If attacks were concentrated solely on cities, exchanges of only a few hundred megatons could lead to a full-scale nuclear winter. At the other extreme, a 2,000 megaton nuclear exchange would probably cause relatively mild changes of climate, if the attacks could be confined to missile silos and airfields far from urban areas. It seems clear, however, that the risk of nuclear winter would be very substantially reduced if the world's nuclear stockpiles could be cut to less than about 2,000 warheads, with a combined yield of no more than 500–2,000 megatons. This would involve the removal of at least 95 per cent of the present nuclear stockpiles. Since the economy and the society of both the United States and the Soviet Union could be devastated with only a few hundred nuclear warheads, it might therefore seem possible to maintain credible nuclear deterrent postures even at these relatively low levels.

For Sagan and many in the arms control community or anti-nuclear movement this argument for radical reduction in the world's nuclear arsenals was logical and persuasive. Had the possibility of climatic catastrophe been appreciated in the mid-1950s, before nuclear arsenals had increased beyond the threshold level, the defence community and the United States government itself may have been more impressed with the argument. It is normally easier to reject future military options than it is to dismantle a vast complex of well-established weapons systems and military institutions.

In practice, however, Sagan's advice did not appeal to much of the American defence community. A range of arguments was put forward by the defence community against Sagan's proposal for deep cuts.[45]. Such cuts would necessarily involve a complete reappraisal of existing strategies and targeting plans. There would simply not be enough warheads to cover existing target sets (the existing American targeting plans include some 40,000 targets) and many existing counterforce options would have to be abandoned. The policy of extended deterrence would also be undermined. NATO at present deploys more than 4,500 nuclear warheads in and around Europe alone, and cuts to below a nuclear winter threshold would involve the removal of almost all of these. The American commitment to use nuclear weapons in response to a successful Warsaw Pact advance on West Germany (the credibility of which is already widely questioned) would arguably be completely undermined, with potentially major effects on NATO as an alliance. That is, for many supporters of present policies, Sagan's proposals would increase the risk of war. This argument does not deal with the extraordinary *global* risks posed by nuclear winter but it does make sense in terms of the *United States'* national security, since American society could be threatened with destruction even if nuclear stockpiles were reduced to the levels Sagan suggests.

The Washington strategist Keith Payne[46] argued that the implications of Sagan's argument are in any case unclear for a United States government. The superpowers had regularly claimed that they wanted to negotiate major cuts in their nuclear arsenals and yet little had been achieved. Payne found it hard to see how the nuclear winter hypothesis would help to remove the obstacles that had so far blocked progress (which for him mainly involved Soviet intransigence and verification problems). Unilateral cuts by the United States would not deal with the risks of nuclear winter, since the remaining Soviet forces would be far above 'threshold' levels. So, he argued, there was little alternative to

The Risk of Nuclear Winter

continuing with present negotiations, though perhaps with renewed vigour.

If there was disagreement about Sagan's arguments for cuts, there was more general agreement that nuclear winter research had a potentially significant effect on the strategic environment. For example, even a perfectly executed first strike against the adversary's nuclear forces could trigger a nuclear winter which would destroy the attacking side's own society. So the possibility of nuclear winter was an additional disincentive to a premeditated first strike.

However, many people felt that it was not clear whether this extra disincentive would make much difference in practice, as long as both superpowers maintained sufficient nuclear forces for assured retaliation. In fact the threat of a premeditated first strike worries most analysts less than the danger of a pre-emptive attack in a crisis: in the belief that war is imminent, each side may be tempted to strike first as the 'least bad' option. In this case the possibility of nuclear winter could generate additional caution but it could also induce extra stress and fear as the crisis intensified and thus degrade the quality of decision making. Moreover, leaders might still be tempted to pre-empt. They might perceive a choice between only two options: devastation plus possible nuclear winter (if 'they' strike first) or less devastation plus possible nuclear winter (if 'we' strike first).

One group in the United States might have been expected to find the strategic implications of the risk of nuclear winter particularly significant. Many on the political right, including Professor Richard Pipes and other past members of the Committee on the Present Danger, have feared that Soviet leaders may believe that their society could survive a nuclear war with losses of 20–30 million, which Pipes claims the Soviet leadership would deem to be an acceptable price for the destruction of the United States.[47] The possibility of nuclear winter would surely change the strategic calculations of the most ruthless conceivable Soviet leaders.

In fact, however, people holding such extreme fears about Soviet intentions were unconvinced about the beneficial effects nuclear winter rsearch might have on Soviet behaviour. Instead they perceived a new threat—a possible 'perception gap'. Cresson Kearny, a retired engineer from Oak Ridge National Laboratory, argued that 'the most destabilizing strategic situation that might result from the "nuclear winter" controversy would be for American leaders to believe that a nuclear war will result in catastrophic climatic effects, while Soviet leaders believe that "nuclear winter" is a myth or a very unlikely possibility.'[48] This could lead to self-deterrence which could be exploited by the Soviet Union: 'Helping to increase fears of unsurvivable "nuclear winter" is an effective means by which the Soviet Union advances its world-dominating objectives while incurring a small risk of nuclear war'.[49]

The danger of such an 'asymmetry of perceptions' was much discussed by the political right and raised repeatedly by Belt-way bandits and other defence analysts.[50] The conclusion which they, and Cresson Kearny, drew was that the United States should not, as might appear logical from their point of view, promote international scientific exchange of information and try to persuade Soviet leaders of the potential dangers. Instead they recommended the United States government to downplay the risks and act as if nuclear winter were extremely unlikely.

Once the United States defence community began to elaborate on the detailed strategic implications of nuclear winter, discussion became predictably arcane. Close consideration was given to the way in which any nuclear winter 'threshold' might be manipulated in nuclear warfighting plans. For example, one side might launch an attack just below the 'threshold', leaving the victim with a choice between surrender or triggering nuclear winter. Undeterred by scientific advice that there was no precise nuclear winter threshold and that, even if there were, we would never know what it was, many analysts used the nuclear winter hypothesis to construct a range of new nuclear warfighting scenarios.[51]

Numerous technical reforms of American defence policy were proposed to reduce the risk of nuclear winter. Measures were proposed that would tend to reduce the extent of fires in urban areas, with their associated smoke and dust. Edward Teller, former director of the Lawrence Livermore nuclear weapons laboratory, argued that the yields of nuclear warheads should be reduced and new earth-penetrating warheads should be widely introduced. Such low yield, earth-penetrating warheads would each ignite fires over smaller areas (a few rather than hundreds of square kilometres).

As it happened, these warheads were being developed in the nuclear weapons laboratories in which Teller worked. Large numbers of such warheads could be combined with new high-accuracy missiles, such as the Pershing II, to attack military targets in the Soviet Union with much less risk of major climatic catastrophe.[52]

Supporters of counterforce doctrines and limited nuclear options argued that the risk of nuclear winter increased the importance of avoiding the targeting of cities and of preparing for limited nuclear war rather than simply for mutual assured destruction. At present American nuclear forces are, apparently, not targeted on cities *per se*. Nevertheless there are 60 targets within Moscow alone and plans for major counterforce attacks would involve the destruction of most large Soviet cities.[53] Thus some analysts hoped that the nuclear winter debates would stimulate the United States military to revise their plans to avoid cities, even though there is serious doubt that this could be done without undermining the military coherence of many counterforce attacks.[54]

The list of such reforms is potentially endless. The possibility was raised that potential targets should be relocated, either towards or away from cities, according to whether the proposer preferred mutual assured destruction or limited nuclear options policies.[55] There was discussion of the implications of any cloud of smoke and dust for intra-war satellite surveillance and communications, or for the operation of aircraft or missiles as they passed through the atmosphere. The balance of the effect of the nuclear winter hypothesis on American preparations for a prolonged nuclear war was weighed:[56] the smoke cloud would make the maintenance of command and control much harder; but if the nuclear attacks could be spread over several weeks, the climatic effects could be reduced.[56] A RAND analyst even considered whether nuclear winter could not provide fanatical regimes with a new terrorist tool: they could threaten to blow their country up or set it on fire in order to devastate the USA through climatic change.[58] Not surprisingly, advocates of the strategic defense initiative (SDI) also found support in nuclear winter for their ideas. The SDI's ballistic missile defences would aim to intercept warheads far above the ground, preventing them from starting fires.[59]

It is hard to gauge how seriously some of these arguments were taken, even by their authors. In the main, nuclear winter seemed to be used to justify changes that people already supported for other reasons. The research also acted as a focus for contradictions or dilemmas that would have existed without the risk of nuclear winter, and sometimes simply as an occasion on which to write a report.

For supporters of 95 per cent cuts none of the implications raised by the defence community addressed the central issues of risk. Introducing new weapons systems in the United States could not remove the risk of nuclear winter unless the existing large yield and non-earth-penetrating warheads were scrapped and unless similar measures were adopted by the Soviet Union, France, the United Kingdom and China. Limited nuclear options and changes in targeting plans would have little effect on risks unless there were a guarantee that they would be adhered to in war and that the war would not escalate.[60]

Several commentators doubted that nuclear winter research should or would have *any* significant security implications. The uncertainties were claimed to be too great for any coherent fine tuning of targeting policies.[61] It was argued that governments of nuclear weapons states would continue to try to avoid nuclear war since such a war would be disastrous for their own countries, with or without nuclear winter.[62] In practice, the nuclear weapons states would remain unmoved by the newly-apparent risks posed to non-combatant countries. It was argued that, to the extent that governments should react at all to the provisional scientific conclusions about nuclear winter, they should ensure that they would have 'no regrets' later by only implementing changes that would be prudent even if nuclear winter proved to be very unlikely.[63]

All the protagonists were agreed, however, about the potential political significance of the nuclear winter debate itself. It was hoped, or feared that public opinion could force changes in United States policy. But some feared that the Soviet government, being less susceptible to public pressure, might take advantage of this factor. There were fears, at this time of tension within the NATO alliance, that nuclear winter might further erode European governments' confidence in the American nuclear commitment to Europe.[64] As the 1985 Non-Proliferation Treaty review conference approached, it seemed possible that the risk of nuclear winter could strengthen non-nuclear and non-aligned countries in their resolve to harass or embarrass the nuclear weapons states.[65] There were even fears that nuclear winter could level down the significance of the superpower nuclear arsenals: if the nuclear forces of Britain, France, China or even Israel could trigger major climatic changes, the vast superpower arsenals could lose their relative symbolic power.[66] Joseph Nye, a former senior official at the State Department during the Carter administration, and others feared that the United States could become diplomatically isolated, unless the administration at least demonstrated some concern about the risks which nuclear winter posed to the Third World.[67]

In practice, nuclear winter provoked some uncomfortable public debates for the United States administration and stimulated an energetic discussion within and between branches of government, as we shall see in the next section. However, the feared diplomatic repercussions proved to be relatively weak, as we shall discuss later.

Debates within the American Government

The Administration's reaction

The United States administration was admirably alert to the new findings, at least during the early stages of the nuclear winter research. When Department of Defense officials first heard about TTAPS's preliminary results in late 1982, scientists from the Lawrence Livermore laboratory (including M. May and T. Malone) were invited to give an initial assessment. Dr DeLauers, advised that there was a serious possibility of climatic effects, asked the National Academy of Sciences to prepare a report on the question and expanded research funded by the Defence Nuclear Agency in the area. Commodore R. Bacon, Director of the Strategic and Theater Nuclear Warfare Division of the United States Navy, was sufficiently concerned to write in July 1983 to the chairman of the Department of Energy's Military Liaison Committee urging that the Department of Defense should do more work on the issue.[68] The Chief of Naval Operations sent representatives to the Conference on the World after Nuclear War in November 1983. They produced a technical summary of the proceedings and stimulated a memorandum suggesting a new in-house study of the potential implications for targeting.[69] One representative also prepared a memorandum examining the potential themes nuclear winter might offer to Soviet propagandists (and Freeze activists) and listing a series of counterarguments.[70]

The public debate that followed the 1983 conference compelled politicians and senior defence officials to make position statements on nuclear winter issues. However, it was a series of congressional and Senate hearings that provided a forum for sustained debate about the implications for American policy. These hearings, and congressional committee requests for documents from the Department of Defense, forced the administration publically to address the issues. The most important hearings were held in July 1984,[71] September 1984,[72] March 1985,[73] and October 1985.[74] There were also important congressional debates and resolutions, requiring the Pentagon to produce, under the Department of Defense Authorization Act, a comprehensive study of the atmospheric and environmental effects of nuclear war and its implications for nuclear weapons strategy and policy, arms control policy, and the civil defence policy of the United States.

This Pentagon study became a focus for congressional pressure on the Department of Defense. When the first report was presented in March 1985, Congress was less than impressed. Their requested comprehensive study amounted to 17 typewritten pages.[75] It said that, although nuclear winter was a possibility, the uncertainties were great and the policy implications, inasmuch as they existed, were that the administration's existing programmes should be maintained or strengthened.

On 2 and 3 April 1985 respectively, motions were introduced in the Senate and the House of Representatives requiring the Pentagon to produce another, better report. In 1986 a report of five pages was presented.[76] Congressional representatives and senators railed against the Pentagon, but so far have not succeeded in eliciting any more detailed response. In the meantime Congress commissioned a similar report from the General Accounting Office. This report discussed the issues in much more detail.[77]

Congressional hearings: a closer look at the debate

We finish this section by considering the way in which scientists, experts and policy makers interacted in one of the Senate hearings — those of the Committee on Armed Services on 2–3 October 1985.[78]

The first of the two hearings dealt with mainly technical issues. It addressed the current scientific findings on nuclear winter and related phenomena; gave testimony on the targeting policy implications; and indicated Soviet contributions to and propaganda use of the nuclear winter hypothesis. The second hearing was supposed to explore the implications of current scientific findings for current nuclear policies. As the debate took place, however, these distinctions were blurred; issues of scientific validity were raised in the second session and policy questions were considered in the first session.

Expert witnesses at the hearings included: (*i*) scientists who had been involved in nuclear winter research, represented by Drs Mark Harwell and Carl Sagan of Cornell University; (*ii*) representatives of the American defence establishment, Dr C. M. Gillespie (Defense Nuclear Agency) and Richard Perle (Department of Defense); and (*iii*) additional witnesses, Leon Sloss (consultant in American targeting policy) and Dr Leon Goure (Centre for Soviet Studies, Science Applications International Corporation). The Senators taking part were Barry Goldwater (Chairman), William S. Cohen, Jeremiah Denton, Alan J. Dixon, J. James Exon, John Glenn, Gary Hart, Gordon J. Humphrey, Carl Levin, Sam Nunn, Dan Quayle, John W. Warner and Pete Wilson.

The Senators asked each witness to present his case and then asked questions. There was little interaction among the witnesses, except towards the end of the hearings when a heated exchange took place between Carl Sagan and Richard Perle. As discussed earlier, the tone of such hearings is overtly scientific, but witnesses and committee members are not entirely disinterested. They are involved in a *political* debate in which skill in scoring debating points and the discrediting of the opposition may be just as important as the intrinsic merit of the arguments.

The language used in the debates is analysed here to indicate how science and policy interacted in the debate about the risks of nuclear winter. The quotations reproduced here, being verbatim reports of spoken contributions, are often less carefully constructed than a written statement, but this correspondingly increases their value for our purposes. Unless otherwise stated, parentheses and emphases in quotations from the hearings have been added by us.

Most participants referred to their scientific credentials or to the place of science in the debate. Expert witnesses were at pains to emphasise their fairness and impartiality. Politicians stressed the importance of science in the policy-making process, the scientific origins of their interest in the subject, or apologised for their personal lack of scientific expertise.

The politicians' introductory statements illustrate these points. For example, Barry Goldwater began by saying, 'I do believe in the scientific possibility of nuclear winter'.[79] He emphasised the importance of the hearings in terms of public information: 'I believe it is important to have hearings like this so that the public, as well as Senators, can be informed about the state of science regarding these issues'.[80] James Exon drew attention to the fact that he had 'attended several conferences in this subject'[81] while John Glenn claimed an interest in the impact of science on policy going back to Van Allen's study of the radiation belts around the globe.

Senator Goldwater, in the chair, referred in both sessions to what he regarded as a 'scientific fact':

> I have spent two wonderful days with Dr Alvarez, the Nobel physicist, discussing the Nemesis theory of what happened to the dinosaurs. *The fact* that every 26 million years the elliptical orbiting of the sun by a star that up to now has not been located fascinated me. I am a firm believer that there could be such a thing as a nuclear winter just as we *had* a winter caused by a portion of a planet hitting this earth.[82]
>
> ... *the fact* that there is that possibility of killing off a lot of people like the Nemesis theory killed off all the dinosaurs and I do not want to be one of those dinosaurs. That is one of the reasons for these hearings.[83]

The Nemesis theory is still the subject of considerable debate in the scientific community and is possibly rather less well founded than the nuclear winter theory itself. However, Carl Sagan refrained from mentioning this in his reply:

> There is now a very widely accepted view, tracing directly from the discovery of Dr Alvarez and his collaborators, that 65 million years ago there was an impact ... [and that] the earth was darkened and cold sufficiently to produce massive extinction. All the species of dinosaurs were wiped out certainly, but in addition most of the species of life on earth were rendered extinct.[84]

The expert witnesses, describing their roles in the hearings, began by defining their areas of expertise and emphasising their impartiality. The defence establishment experts introduced themselves as follows:

> ... the role of the Defense Nuclear Agency in this matter is to manage scientific research into the climatological effects associated with what we call nuclear winter. As such, then we have no position on the science of nuclear winter, much less on the policy implications that flow from those science considerations .... I have tried to make it as nearly the general consensus of the scientific community as I am able to do that. [C. M. Gillespie[85]]
>
> ... those of us who are responsible for developing and managing American strategic policy are indebted to the community of scientists who have alerted us to the danger that a nuclear war could lead to profoundly destructive climatic

change. But . . . its contribution to the wisdom of our thinking about war and peace is far from clear. [R. Perle[86]]

From the additional independent experts we have the following:

> . . . although I think I know enough from careful technical studies about the phenomenon . . . of nuclear winter, I will focus my remarks on policy issues with particular emphasis on nuclear strategy and targeting strategies and weapon results. [L. Sloss[87]]
>
> I must emphasise that my remarks are based entirely on personal views and on personal research. I have been a long-time student of Soviet views on the consequences of nuclear attacks and also of measures to mitigate its consequences. [L. Goure[88]]

Mark Harwell, one of the nuclear winter scientists, gave a one-page account of the SCOPE project and his role in it and summarised as follows:

> I . . . consider myself as spokesperson for the SCOPE project . . . [which] involved something like 300 scientists from over 30 countries around the world trying to assess, based on existing data and existing models, what the consequences of a nuclear war would be.[89]

Carl Sagan made no such initial statement before beginning a detailed description of the consequences of nuclear war and the scientific support for the nuclear winter hypothesis.

The scene was therefore set by most senators and witnesses in terms that gave primacy of place to the role of science in the policy area. Senator Cohen was the only one to attempt openly to put science on an equal footing alongside the political process, and it is interesting to note that, of all the senators, he seemed to have the best understanding of the scientific aspects of the subject. Richard Perle paid lip service to the role of science but then cast doubt on its place in the policy-making process.

A reverential but incomprehensive attitude to science was evident in many of the questions asked by the senators of the expert witnesses. Gary Hart said in his opening statement, 'Those of us who are not scientists worship the precision of science, the certitude of science.'[90]

As the hearings proceeded, it became clear that the nuclear winter scientists profoundly disagreed with the other witnesses about whether the possibility of nuclear winter should lead to a major reassessment of the risks associated with nuclear deterrence or whether it was merely a minor consideration, implying no more than fine adjustments to present policies. A number of social scientists, including Nelkin,[91] have studied risk-related conflicts. Nelkin has described how the choice of language used in debates such as the Senate hearings has both judgemental and strategic implications. It is judgemental in that the selective use of particular labels can either trivialise or render important an event or concept; marginalise or empower groups or individuals;

The Risk of Nuclear Winter

reduce issues to routine or define them as problems. It is strategic in that it carries implications for the formulation of policy. The conflict described here was reflected in the judgemental language witnesses used in arguing their case. Witnesses engaged in the selective use of labels to devalue the evidence presented by others with whom they disagreed. Those who denied that nuclear winter had any significance would use language aimed to trivialise concepts or evidence supporting the nuclear winter hypothesis; to marginalise groups or individuals involved in nuclear winter research; and to reduce nuclear winter issues to the routine. Supporters of the nuclear winter hypothesis would use language aiming to do the reverse.

Having established in his opening statement that he would try to make his testimony as close to the general scientific consensus as possible, Gillespie said that the TTAPS results were 'upper limits' and that a severe nuclear winter was extremely unlikely. While this was quite possibly true, it did not represent the scientific consensus at the time (although the word 'severe' may, of course, be interpreted in various ways). Of Gillespie's judgemental statements, nine used language which tended to trivialise the concepts or events that support the nuclear winter concept and five tended to depict nuclear winter issues as trivial or routine.[92] This is illustrated by statements such as:

> All of these studies considered a *so-called* baseline case . . . the baseline nuclear winter effects are as predicted about as bad as they can get.[93]
>
> The initial assertion that that [eradication of the human species] was a distinct possibility was not based upon any very careful or thoughtful assessment.[94]
>
> It does appear now that we do recover from the nuclear winter over a period of weeks or months or in some cases several years perhaps, depending on how certain parameters go.[95]
>
> Give their assumptions, I or anyone else would have arrived at the same conclusion . . . the question is, are all their assumptions . . . all right?[96]
>
> [In response to Harwell's claim that the conclusions of the modelling exercise were robust to changing scenarios] . . . the terrible word robust conclusion was just used . . . *they changed the ground rules*. They increased the assumption as to the amount of smoke produced in order to keep the total amount injected roughly the same.[97]
>
> . . . you see, nothing is specified, how much of them do you burn? What cities? Do you assume that a nuclear weapon is just a match that starts the whole thing burning, or do you more appropriately perhaps relate the amount burned to the yield of the nuclear device? I find the statement that burning some handful of cities is almost sure to lead to the climatic catastrophe a rather suspect statement . . .[98]

Richard Pearle's short statement pointed out that the aim of strategic policy was to prevent war and that this would be accomplished by maintaining a stable balance of nuclear forces. This situation would not be materially affected by the discovery of nuclear winter. He concluded by saying:

> The one [thing] I have trouble understanding is a mindless embrace of the notion that we should freeze our nuclear forces, abandon research on strategic defenses and reaffirm the policy of attacking urban centres in the Soviet Union by proclaiming the tonic effect of mutual assured destruction.[99]

Perle is a robust debater and several of his statements were concerned to trivialise, marginalise, or reduce to the routine the nuclear winter research. Examples are as follows:

> I was grateful to Senator Nunn for a question that caused Dr Sagan to be precise about the conditions. . . . Too often there is a casual transfer of statements that we can make with confidence into areas where we can only speak with less confidence. . . . It is important for the public to recognize where science can make precise judgements and where it cannot.[100]
>
> [In relation to a diagram produced by Sagan] What I find uninformative about a chart like that, although it is a *very attractive piece of artwork*, is the notion that lies behind it of the Soviets launching 10,000 nuclear weapons. . . . It strikes me as such a preposterous assumption that any chart that purports to describe the consequences of it seems to me to miss the point entirely.[101]

Leon Sloss argued that nuclear winter had been 'both exaggerated and underestimated in importance'. He went on to explain why he felt that the scenarios were unrealistic, emphasising that the nuclear winter phenomenon was likely to be inconsequential in relation to the direct effects of blast and fallout. He felt that analyses should concentrate on 'mild nuclear winters': limited nuclear war was a strong possibility 'because all parties involved will have a strong interest in keeping it limited'. He went on to suggest a range of policy options aimed at minimising the climatic effects of a limited nuclear exchange. Sloss mainly contributed to the debate on policy matters, but commented on the validity of the TTAPS study that:

> It is hard to be critical of the scenarios because they have a whole *slew* of scenarios. . . .[102]

His introductory analysis seemed to be one long attempt to marginalise or discredit the sources of nuclear winter research. He said later in his testimony:

> In the Soviet Union there is no such thing as non-government research, non-government approved writings. . . .[103]
>
> One of the problems, I think, is that many of the images we have of the nuclear winter effect generally seems to assume that the detonation is always downtown of every large city. Why anybody goes out to hit city hall I don't know. From a military point of view, that targeting is absurd.[104]
>
> From the point of view of the belligerent countries . . . a nuclear winter effect of a few weeks or a month is irrelevant. It is hell for other people, but for people subjected to this, this is the least of their worries.[105]

All fair statements, perhaps, but said with a view to trivialise or marginalise the nuclear winter issue and misrepresent his critics' arguments.

The scientists involved in nuclear winter research tended to make less use of overtly judgemental language. They were also involved in advocacy, but their presentation of the significance of nuclear winter was milder and more qualified. Their language tended to be less dismissive of the arguments of their critics. This may have been partly because they believed that all they had to do was to establish in the minds of senators that there was a substantial risk of nuclear winter — a case for which the scientific evidence was very strong. Thus they could afford the luxury of occupying the high ground of debating style.

Harwell described how the uncertainties in many of the earlier modelling studies had been resolved, but indicated that this had created fresh areas of uncertainty. Scientists' predictions of ecological, agricultural and human consequences were based on the vulnerability of biological systems to physical stress. These consequences were found to be so great that many of the inherent uncertainties could be dismissed as unimportant. Despite the fact that all basic assumptions overestimated the number of people that could be kept alive, models indicated that the primary mechanism for human fatalities would probably be mass starvation. They also suggested that people in non-combatant countries, in Asia or Africa for example, were at least as much at risk.

His statements were characteristically designed to make the nuclear winter research appear both professional and significant:

> In the TTAPS study there was a large number of scenarios looked at. . . . There were sensitivity analyses of what . . . the effects of changing the various parameters were. It was impressive how robust the answers were to changing scenarios.[106]

Sagan providing additional scientific background to the nuclear winter hypothesis, referred to the fact that more than 50 different scenarios had been tested. Even studies carried out by initially sceptical bodies had concluded that nuclear winter was a clear possibility. He stressed the policy implications and criticised the Department of Defense's claim that the nuclear winter hypothesis strengthened the case for the SDI. In Sagan's view, SDI would not help to prevent a nuclear winter and could increase its probability. His summing up stressed that the occurrence of nuclear winter did not depend on a 'worst case' scenario and that safety could only be ensured by reducing the number of strategic nuclear weapons to a point below which nuclear winter could not be triggered.

Thus Sagan was the strongest expert advocate for the significance of nuclear winter research, and he phrased his arguments in order to make the greatest impact. For example:

> [T]his [initiation of a first strike] is an extremely expensive and elaborate way to commit national suicide.[107]

The above examples and summaries illustrate the clear demarcation between the positive statements used by Harwell and Sagan and the predominantly dismissive or critical language used by all the other witnesses. As noted above, the process was one of advocacy, rather than scientific debate, and this would therefore be the expected pattern. However, the nature of the statements is interesting: many of the critical statements were directed at the scientific basis of the nuclear winter hypothesis. It was legitimate to point to uncertainties but the senators were hardly in a position to assess the scientific merits of the arguments. Since the hearings were, of their nature, before politicians rather than senior scientists, superficial scientific debates became, not unexpectedly, a vehicle for arguments about policy. Perhaps it would have been wiser for the hearings to take the scientific assessment of the NAS and SCOPE reports as a basis for policy debates. But to expect politicians, representatives of interest groups or even concerned scientists to conduct a purely academic analysis in such a forum is unrealistic.

Harwell and Sagan emphasised that exemplary scientific procedures had been used in nuclear winter research: the number of different scenarios that had been tested, the conservative nature of the assumptions made and the detailed sensitivity analyses carried out on them, plus the robustness of their conclusions. However, the other witnesses described these activities as 'changing the ground rules' or adopting 'a whole slew of scenarios'; they focused their attention on one scenario and ignored the others; and they referred to the underlying assumptions as 'preposterous'.

A large number of scientists, worldwide, had contributed to research on nuclear winter, and many of those involved valued their collaboration with scientists in the Soviet Union.[108] Harwell and Sagan referred to the endorsement and support of this wider scientific community, but the other witnesses largely ignored the support of Western scientists and concentrated their attack on the Soviet contribution.

To summarise the value judgements on each side of the conflict: the nuclear winter scientists viewed the world's ecosystems, particularly agricultural systems, as fragile and easily disrupted, and the science underlying their conclusions as basically sound; the other witnesses regarded the science as unsound and emphasised the resilience of climatic systems.

The divergent attitudes to the nuclear winter hypothesis and its significance, noted in the previous section, were reflected in very different views of the relevant policy options. The nuclear winter scientists, perceiving a threat, talked in terms of global risks and of radical changes in the strategic situation. They recommended a reduction in the world stockpile of nuclear weapons to below the number capable of causing nuclear winter as the only rational policy option. On the other hand, defence establishment scientists and some other witnesses acknowledged a smaller risk, that added only marginal extra risks to existing policies. They talked in tactical terms of the avoidance of nuclear war during summer in the northern hemisphere, mutual agreements between the superpowers not to attack major cities and the development of nuclear weapons of a type that would be less likely to start fires.

The nuclear winter debate outside the USA

Outside the United States nuclear winter research has stimulated a range of debates with varying degrees of intensity and official interest. This section briefly describes some of these, to offer some comparison with the situation in the United States.

The United Kingdom

British scientists were involved in the original special edition of *Ambio*, but few played a role in the early stages of nuclear winter research in 1983. It was not until after the Washington conference, which a number of British scientists attended, that the risk of nuclear winter was widely discussed.

The conference took place at a time when the Campaign for Nuclear Disarmament and the broader anti-nuclear movement were at a peak of activity: the first ground-launched cruise missiles were deployed at Greenham Common in November 1983. Nuclear weapons issues had been at the top of the political agenda during the June general election and, although Labour's policy of unilateral British nuclear disarmament had only minority support, public opinion polls clearly indicated widspread concern about the dangers of the nuclear arms race and the risks of nuclear war. Aside from the cruise missile debate, the decision to buy Trident was highly controversial and (as in the USA) there was a major and effective campaign to discredit the effectiveness of the government's civil defence programme. In this context the nuclear winter research was bound to stimulate controversy.

The risk of nuclear winter was, in Britain as in the United States, mainly used to reinforce existing policy positions. Denis Healey, the Labour Party's foreign affairs spokesman, quoted the results of some of TTAPS's smaller scale scenarios to emphasise the potential destructiveness of the four Trident submarines ordered by the British government.[109] He simultaneously exploited the possibility of nuclear winter to persuade a section of his unilateralists opponents in the Labour Party that they could not hope to make Britain safe from nuclear attack simply by withdrawing from NATO and expelling American bases. The Campaign for Nuclear Disarmament exploited the risk of climatic catastrophe in the same way: to attack the cruise, Pershing and Trident missile programmes and to emphasise the importance of international nuclear disarmament. However, many in the anti-nuclear movement were nervous of emphasising the nuclear winter findings. There was a fear (also expressed within the American Freeze movement and the arms control community) that if the nuclear winter research were discredited, the public might neglect the other horrors of nuclear war and believe that the dangers to them were exaggerated.[110]

There were, however, significant differences between the British and the American debate. Scientists in both countries featured in television and radio programmes and acted as consultants; but British scientists have access to fewer established fora for participation in public policy debates. No parliamentary committee hearings dealt with the nuclear winter issue.

The British scientists' organisation which did the most in the early stages to draw attention to the risk of nuclear winter—Scientists Against Nuclear Arms (SANA)—was a group committed to raising awareness of the risks of nuclear deterrence. Its members wrote books, articles in a wide range of magazines and contributed to radio and television programmes. SANA issued a pamphlet and a video cassette on the nuclear winter issue, organised a number of scientific review meetings and public conferences and arranged in November 1984 a major tour of Britain for four key American and Soviet scientists involved in nuclear winter research: Richard Turco, Paul and Anne Ehrlich and Georgiy Golitsyn.[111] A number of SANA members also contributed in a professional capacity to the SCOPE report and review process, but other British scientists connected with SCOPE were reluctant to participate in the politically charged public debate.

The British government maintained that the uncertainties involved in the nuclear winter research were too great to warrant any policy review. There was no mechanism in Britain, similar to those of the American Congress, to force the Government to address the issues. Uncertainty was used as an excuse to procrastinate. The Government, replying in 1984 to a parliamentary question on the implications of nuclear winter for civil defence preparation, said that it was waiting for the results of the SCOPE report.[112] After the SCOPE report was published, the Government still refused to make a statement on the policy implications of the research.

British debates about the nuclear winter research focused mainly on its implications for civil defence policy. The role of local authorities in civil defence provision meant that this issue could be officially raised by councils critical of civil defence preparations for nuclear war (the local nuclear free zones movement). The Home Office F6 Division and the Scientific Research and Development Branch (SRDB) are, together with a voluntary network of regional scientific advisers, responsible for providing civil defence planners with relevant scientific advice. It is indicative of the Home Office's attitude that the information about nuclear winter research which it chose to circulate to emergency planners was an article by Cresson Kearny, the retired Oak Ridge engineer referred to earlier. This article, which had never been published in a refereed journal, criticised some of the TTAPS assumptions in an idiosyncratic way.[113] The Home Office did not circulate to civil defence staff either the TTAPS article or any other work by scientists involved in nuclear winter research. When Richard Turco and his three colleagues toured Britain and offered briefings to emergency planning staff, the Home Office circulated a letter advising such staff not to attend them.[114]

A broader spectrum of scientists was prepared to contribute to public debate about nuclear winter in Britain after the SCOPE report was published. However, the British policy debate never achieved the intensity or level of detail of that in the United States. The risk of nuclear winter may have been widely established in the mind of the general public but there is no evidence that it has led to any changes of policy.

The Soviet Union

Carl Sagan and his colleagues, as we have seen, involved Soviet scientists in nuclear winter research at a very early stage. By the time of the Washington conference, a number of Soviet scientists had achieved results that qualitatively corroborated the findings of scientists in the United States and Western Europe. Senior academicians, such as Vice President Yevgeny Velikhov, soon became publicly convinced of the seriousness of the risk. The Soviet Union hosted an international scientific conference on nuclear winter in Leningrad in May 1984 and Soviet scientists have continued to participate in nuclear winter research, though with much smaller resources than their Western colleagues.

There appears to be little reason to doubt that the scientific academies and academicians are advising Soviet leaders and officials that nuclear war could lead to a nuclear winter which would ensure the destruction of Soviet society. Velikhov told Sagan in 1983 that 'he had held long discussions with both Foreign Minister Gromyko and with Defence Minister Ustinov about the nuclear winter results.[115] Gromyko is said to have raised the subject at a meeting with the Italian President Andreotti. Velikhov is in regular contact with Gorbachev, and it is widely believed that he acts as the Soviet leader's personal scientific adviser.

This suggests that the former American Defense Secretary Caspar Weinberger's view that the Soviet contribution to nuclear winter research 'has been to mirror back to us our own technical analysis and to exploit the matter for propaganda'[116] cannot be the whole truth. However, it does appear to be true that, at least initially, the Soviet Union concentrated more resources on informing overseas audiences of the risk of nuclear war than on educating the Soviet public about these risks. According to Steven Shenfield, an expert on Soviet policy debates, effective coverage of nuclear winter in the Soviet Union was restricted to the specialist scientific press until the second half of 1984.[117]

In 1984 a 'summary document' circulating in the West purported to show that the Soviet Union had given extensive internal coverage to the nuclear winter issue. This was widely quoted to counter the accusations of the political right in the United States and Britain, and to create a myth among Western liberals that the Soviet authorities kept their own population fully informed about these matters. However, according to Shenfield, this document, the origin of which is unknown, contained significant exaggerations. It is apparently not true that the proceedings of a May 1983 conference in Moscow on the effect of nuclear war was covered by the Soviet mass media in 'great breadth and detail'. Apparently this was only true of the coverage in *The Courier* of the Soviet Academy of Sciences (print run 4,500). Similarly, Soviet coverage of the November 1983 Washington conference was poor. The US–Soviet teleconference was not televised as a whole as in the United States; only brief fragments were shown on one evening news programme. *Pravda* reported only that there has been 'a Soviet–American symposium in Washington on the problems of assessing the global consequences of nuclear war.'[118] According to Shenfield, only a few brief articles on the subject were published before 25 July 1984. It was only then that extensive press coverage was achieved, with a long article by Academician Gol'danskii and Professor Kapitsa in *Izvestia*.

However, internal Soviet treatment of the nuclear winter has apparently become much more extensive since mid-1984. For example, several articles appeared in the June 1985 issue of the popular science journal *Priroda* and two articles appeared in the *Literary Gazette* in winter 1985–86.[119] As Gorbachev's 'glasnost' policy has taken effect, the domestic coverage has become wider still.

Thus it appears that the Soviet leadership was much more hesitant about publicising the risk of nuclear winter than many liberal commentators and writers in the West suggested in 1983–85. But there is evidence of a change of attitude and policy amongst the Soviet elite, suggesting at least a period of serious deliberation on the scientific status of the hypothesis and on its potential policy implications. However even now, the only Soviet book published on the nuclear winter, *The Night After: Climatic and Biological Consequences of a Nuclear War*,[120] is primarily aimed at and distributed among foreign audiences.

Many Soviet citizens may still be receiving contradictory messages about the survivability of nuclear war from official publications. For example, at about the same time as *Priroda* (print run 60,000 copies) published its series of nuclear winter articles, a civil defence textbook (print run 100,000) was published and included statements such as:

> to attain victory in a military conflict, alongside decisive actions by the Armed Forces to repulse attack and rout the adversary, timely preparation to defend the population and the national economy from weapons of mass destruction has enormous significance.[121]

However, virtually all anecdotal and opinion poll evidence about the beliefs of the Soviet public seems to indicate an overwhelming belief that there would be no victors in a nuclear war[122] and that there is widespread cynicism about the effectiveness of civil defence preparations.

When the risk of nuclear winter was first recognised in 1983, the Soviet leaders had already been debating amongst themselves for more than a decade whether nuclear war was survivable. The Soviet leaders Malenkov and Kruschev periodically argued, as early as the 1950s and 1960s, that a nuclear war *could* destroy world civilisation. This argument rose and fell according to the tactics of their own power struggle. It was also used in Kruschev's bid to reform the armed forces, and as part of Soviet polemics against Maoist China.[123]

However, the prospect of survival and victory was not repudiated, and it was not until the early 1970s that Brezhnev and his associates consistently referred in public to nuclear war as a *threat* to global civilisation.[124] It appears that, behind this public position, an intense internal debate has been conducted within the Soviet leadership about the *inevitability* of destruction after a nuclear war and the utility of pre-emptive strikes to reduce the scale of attacks on Soviet territory. Most Western analysts agree that the Soviet military tend to stress that global destruction is only a possibility, leaving a role for Soviet traditional nuclear strategy and preparations for prompt military action. In contrast, many scientists and civilians have argued for the virtual inevitability of total annihilation in the event of nuclear war.

The risk of nuclear winter seems to have entered this debate at a time when those arguing for the impossibility of victory already seemed to be winning the argument. The new risk has been an important and often quoted factor in emphasising the importance of 'new thinking' about the security situation. This new thinking has, under Gorbachev, been advocated at the highest levels. Yakovlev, one of the fastest rising stars in the Politburo, has said, 'Sometimes, the thesis that there cannot be a winner in a nuclear war is considered a pacifist formula. Specialists in physics, chemistry, biology and medicine... must give this formula the character of a universally recognised and indisputable truth'.[125] Gorbachev stated on Soviet television in August 1986 that the nuclear winter findings showed that nuclear attackers would be committing suicide as a consequence, 'not even of the counterstrike, but of the consequences of the explosion of his own warheads'.[126] He has since repeated this argument, notably at the Moscow Peace Forum in February 1987.[127]

Western fears that the Soviet leadership may privately discount the risk of nuclear winter, and only refer to it for propaganda purposes, seem therefore unfounded. Indeed to the extent that any 'perception gap' exists, it could operate in the opposite direction from that feared by Washington's Belt-way bandits. The Soviet leadership may well be more conscious of the risks of nuclear winter than the Reagan administration, which has tended to point with relief to any calculation indicating that the climatic effects of nuclear war could be less than is suggested in the SCOPE report. Furthermore, it is widely recognised that the Soviet Union would be more vulnerable than the United States to even relatively mild climatic changes. This is a consequence of its more northerly latitudes and the problems with Soviet agricultural production.

Although the Soviet Union has exploited the risk of nuclear winter in its foreign propaganda, it is not clear that the hypothesis is quite the bonus for the Soviet Union which the Reagan administration and others in the West have supposed. Soviet leaders have promised not to target non-nuclear countries which have no American nor NATO bases, and Soviet propaganda has repeatedly reminded countries with American bases that they are making themselves a target. A nuclear winter would affect countries irrespective of whether they were directly attacked, a point which undermines this long-established Soviet propaganda argument.

There has been no public discussion in the Soviet Union of the implications of nuclear winter for detailed targeting, arms development and procurement policy. It is impossible to tell whether the Soviet military have altered their operational plans in response to the nuclear winter research, but the Soviet military and the defence community, like their American counterparts, appeared to be little impressed by the research. Thus the military changes have probably been minimal in this country where the implementation of defence preparation is traditionally left to the military.

This probability would cause little concern to those like Sagan, who believe that changes in targeting or deployments would be an inadequate response to the risk of nuclear winter, especially in view of the likely overall changes in attitudes to nuclear war amongst the Soviet leadership. However, some Western defence analysts might find it more discouraging. There is little point in changing American targeting plans if any Soviet response would continue to pose a great risk of triggering nuclear winter.

Outside NATO and the Warsaw Pact

The nuclear winter research suggested that, after a major nuclear war, the number of casualties outside the nuclear alliances could greatly exceed the number of deaths and injuries within NATO and the Warsaw Pact. Hundreds of millions of people would die within Europe, the Soviet Union and the United States but billions could die elsewhere. It is therefore not surprising that the potential global consequences of nuclear war have attracted attention worldwide. Many commentators believe that the most important political consequence of nuclear winter research could arise from reactions in non-aligned or non-nuclear countries.[128]

The research has certainly had an effect at the diplomatic level. Nuclear winter has been discussed at the United Nations General Assembly, which asked the Secretary-General to compile and distribute a dossier of scientific studies on nuclear winter and urged all states and international organisations to co-operate in its compilation and distribution.[129] This resolution was passed on 17 December 1984 by 130 votes to nil, with 11 abstentions from the main NATO states, Israel and Colombia. Raoul Alfonsin, Rajiv Gandhi, Miguel de la Madrid, Julius Nyerere, Olaf Palme and Andreas Papandreou, the leaders of Argentina, India, Mexico, Tanzania, Sweden and Greece respectively, emphasised the risk of nuclear winter in their widely publicised *Delhi Declaration* of January 1985, which urged that the nuclear arms race should be reversed.[130] The nuclear winter issue has reinforced international concern about nuclear arms control and disarmament.

However, it is not clear that the risk of nuclear winter has generated as much additional diplomatic pressure on nuclear weapons states as many had initially hoped or feared. In March 1985, for example, Joseph Nye urged the American administration at least to reassure non-aligned states that it was concerned about the risks of nuclear war. He feared that conspicuous disregard for international concern, illustrated by an abstention on the United Nations resolution, would not only needlessly offer the Soviet Union propaganda advantages, but also could disrupt the forthcoming Non-Proliferation Treaty review conference.[131] In practice, the review conference passed off with much less embarrassment for the nuclear powers than Nye had feared, although national representatives occasionally referred to the risk of nuclear winter. Non-aligned countries already active on nuclear issues have added nuclear winter to their list of concerns, but we know of no evidence that the additional risk has as yet stimulated additional states to new activity.

Perhaps the nuclear winter hypothesis may stimulate countries like the People's Republic of China or France to reconsider their refusal to join the Non-Proliferation Treaty and participate in nuclear arms control and disarmament talks. However, it appears that, of the countries outside NATO and the Warsaw Pact, the nuclear winter research has so far had most political impact within New Zealand and Australia. In New Zealand a hard-hitting scientific report of the New Zealand Ecological Society was followed up in 1987 by a report for the New Zealand Royal Society.[132] This report, unlike its counterparts, examined the potential implications of nuclear winter for New Zealand policy on services, civil defence and other such politicised matters. These reports had significant impact, coming as they did during the debate about ANZUS and New Zealand's place in it.

The ruling Labour Party and the country's anti-nuclear movement used the nuclear winter research to emphasise the importance of nuclear issues for New Zealanders and to justify the high-profile, anti-nuclear position taken by the government.[133] The New Zealand government used a proportion of the compensation paid by France for sinking the Greenpeace boat *Rainbow Warrior* to commission a report in 1987 on the scientific and strategic significance of nuclear winter for the country. When it was published the New Zealand Ministry for the Environment produced a leaflet for mass circulation, publicising the report's findings and inviting public comments on it and on areas for further research.[134] In Australia, too, scientists published books and articles on the potential implications of nuclear winter. Their impact on the public debate was significant and the issue was picked up by the anti-nuclear movement. However, the government response was much less noticeable than in New Zealand.

Conclusions

The interactions between recent scientific research into the global consequences of nuclear war and policy debates have, from the beginning, been more complex than some of the myths about the role of science and scientists in policy formulations might suggest. The initial studies were partly stimulated by renewed concern about the nuclear arms race in the early 1980s, and the way the research was developed and presented by scientists was greatly affected by the desire to influence policy debates. Widespread public alarm at the possibility of nuclear war, and the activities of the Campaign for Nuclear Disarmament, the Campaign for European Nuclear Disarmament, the Freeze campaign and other anti-nuclear movements, ensured that nuclear winter research would provoke controversy. In anticipation of this, scientists took great care with the preparation of their early calculations and analyses, and, in particular, with developing a substantial body of prestigious scientific support for the significance of their findings. Many of the scientists involved, far from being the politically disinterested and naive boffins of common image, showed political skill in their use of scientific findings to stimulate and to set the agenda for a policy debate. They exploited their scientific prestige to secure a public hearing. The scientific institutions also played an important role in reviewing the progress of research and ensuring that the risk of nuclear winter could not be dismissed.

The American military and the State Department were quick to recognise the potential importance of the findings, as, apparently, was the Soviet leadership. As soon as the public debate began in November 1983 officials, institutions and pressure groups adopted positions consistent with their established interests. The initial reaction of the American and the British government was to emphasise the scientific uncertainties involved and to attempt to minimise the political impact of the research. Anything that further increased public unease about the risks of nuclear deterrence was unwelcome. The Soviet leadership too was apparently initially reluctant to publicise the research within the USSR. It did, however, feel free to make great use of it with foreign audiences.

Characteristically, the most detailed public debates about the policy implications of nuclear winter research took place in the United States. There the large strategic studies and defence analysis communities and the system of congressional and Senate hearings provided a context in which many analyses and arguments were presented. Scientists and experts contributed substantially to these hearings. However, the scientific uncertainties and the wide range of its potential policy implications often allowed officials and interest groups to use the nuclear winter research to justify their existing policy positions. The public, and, indeed, many policy makers, seemed poorly prepared to cope adequately with issues of scientific uncertainty and risk, and this frequently meant that debates were confused. The congressional hearings, for example, demonstrated how policy debates, however scientific they are in tone, involve struggles to undermine adversaries' credibility or to score debating points. Bad arguments supported by powerful interest groups may prove more influential than good arguments which are uncomfortable to such groups.

However, the implications of nuclear winter findings do seem to be intrinsically ambiguous and dependent, for example, on the moral standpoint one adopts. Many people in the United States, the Soviet Union and the countries allied to each believe that nuclear war would threaten the total destruction of their societies, even without a nuclear winter. The additional threat of nuclear winter does not radically change the risk assessment for people who are primarily concerned with American or Soviet national security. The research further undermines the credibility of civil defence and emphasises the importance of avoiding nuclear war. But the perceived need for nuclear deterrence in order to prevent such a war would remain as strong or as weak as before, and it is perhaps unsurprising that people taking this value position are reluctant to consider more than an adjustment to existing policies.

Civilian scientists, international organisations, non-aligned countries and the anti-nuclear movements are more inclined to take a global view of the risks. From this point of view the risk of nuclear winter does pose major new threats and demands radical action. Sagan's argument for dismantling the potential doomsday machine by reducing world arsenals by about 95 per cent is persuasive in this context. However, there is still the question of how to combine this with the maintenance of stability; a unilateral act would not tackle the problem at all unless it was reciprocated by other nuclear powers, and multilateral negotiations would confront the same obstacles as before.

From the present authors' point of view, Sagan's argument is persuasive, and proposals to change weapons procurement or targeting plans do not adequately respond to the risks. It is to be hoped that the risk of nuclear winter could at least persuade the governments of nuclear weapons states of the desirability and urgency of major cuts in nuclear arms, opening the way to unilateral initiatives and genuine negotiations. However, it is much harder to dismantle existing systems, with the vested interests that have built up around them, than it is to block future options. Had the risk of nuclear winter been identified in the mid 1950s, while nuclear arsenals were still relatively small, it is possible that arguments against any further expansion of the nuclear arsenals beyond a total of 500–2000 megatons would have prevailed. Strategic thinking about the requirements for nuclear deterrence could have continued to work on the assumption of small nuclear forces. The intellectual and moral arguments against risking the destruction of global civilisation and most of the human race remain just as strong in the 1980s as in the 1950s, but the possibility that they would be willingly accepted by national security and defence communities is now much more remote.

Nuclear winter research exposes and intensifies contradictions in present targeting plans and deterrence policies. However, so many contradictions and confusions exist in the policies already that there is little reason to suppose that the additional findings of nuclear winter researchers will automatically result in any changes.[135] Major changes will come about only as a result of new political pressures. Many nuclear winter scientists and their colleagues recognised this and worked hard to ensure that their findings had a wide impact outside the existing policy makers and defence communities. Four years after the public launch of the nuclear winter research in October–November 1983, it is clear that the risk of nuclear winter has been widely established as an important factor in the debate about nuclear policy and in the minds of world leaders and the general public. Yet there is little evidence that it has resulted in any specific changes in policy.

Perhaps this was always too much to expect. But the nuclear winter hypothesis may have contributed substantially to significant political changes that have occurred since 1983. In the early 1980s senior officials in the Reagan administration were on record as believing that nuclear war would be survivable for the United States.[136] A powerful political movement of the right in America and Britain re-emphasised their faith in civil defence against nuclear attack and challenged the mainstream beliefs of the 1970s about the unwinnability of any nuclear war. By 1987 this powerful political challenge had almost entirely vanished. Right wing rhetoric changed radically as a result of the widespread public debates about nuclear weapons in the 1980s. There may now be a broader consensus than ever before about the unsurvivability of nuclear war. Scientific work on the direct effects of nuclear war and its possible global consequences had a major impact on public debate.

In the Soviet Union too, nuclear winter research may have helped to make the unwinnability of nuclear war the dominant belief of the leadership. Gorbachev and his supporters have frequently referred to nuclear winter to emphasise their 'new thinking' about national security. The concept of 'common security' was formulated before the risk of nuclear winter was identified and was popularised by the Brandt Commission in 1982. The nuclear winter hypothesis can only have helped to persuade people of the importance of this concept in the nuclear age and, in particular, to alert those living outside NATO and Warsaw Pact countries to the fact that it also is relevant to them.

Notes

1. B. Fischhoff, S. Lichtenstein, P. Slovic, S. I. Derby and R. L. Keeney, *Acceptable Risk*, Cambridge University Press, 1981; M. Douglas and A. Wildavsky, *Risk and Culture*, University of California Press, 1982.

2. *New Scientist*, 5 June 1986, p. 21.
3. *Risk Assessment: a Study Group Report*, The Royal Society, London, 1983, pp. 89–91.
4. D. G. Hoel, R. A. Merrill and R. P. Perera (eds.), *Risk Quantification and Regulatory Policy*, 19th Banbury Report, Cold Spring Harbor Laboratory, New York, 1985.
5. M. Rein, 'Value critical policy analysis', in D. Callahan and B. Jennings (eds.), *Ethics, the Social Sciences and Policy Analysis*, Plenum Press, New York, 1983, pp. 83–111.
6. R. L. Meehan, *The Atom and the Fault*, MIT Press, Cambridge, Mass., 1984.
7. Scientific Committee on the Problems of the Environment (SCOPE), *Environmental Consequences of Nuclear War* (2 Volumes), John Wiley, New York, 1985; National Reearch Council of the National Academy of Sciences, Committee on the Atmospheric Effects of Nuclear Explosions, *The Effects on the Atmosphere of a Major Nuclear Exchange*, National Academy Press, Washington, DC, 1985.
8. P. J. Crutzen and J. W. Birks, 'The atmosphere after a nuclear war: twilight at noon', *Ambio*, vol. XI, no. 2–3, 1982, pp. 114–25.
9. P. J. Crutzen, private communication.
10. J. Rotblat, private communication.
11. R. Turco et al., *Long Term Atmospheric and Climatic Consequences of a Nuclear Exchange*, (The Blue Book), Mimeo (127 pp. + maps and figures), March 1983 (available on request from authors); R. Turco, O. Toon, T. Ackerman, J. Pollack and C. Sagan, 'Nuclear winter; global consequences of multiple nuclear explosions', *Science*, vol. 222, pp. 1283–92, 1983.
12. R. Turco, O. Toon, T. Ackerman, J. Pollack and C. Sagan, 1983 (see note 6); P. R. Ehrlich et al., 'Long term biological consequences of nuclear war', *Science*, vol. 222, pp. 1292–1300, 1983.
13. Published later as: C. Covey, S. Schneider and S. Thompson, 'Global atmospheric effects of massive smoke injections from a nuclear war: results from general circulation model simulations', *Nature*, vol. 308, pp. 21–5, 1984; M. MacCracken, 'Nuclear war' preliminary estimates of the climatic effects of a nuclear exchange', paper presented at the 3rd session of the International Seminar on Nuclear War, Sicily, 19–23 August, 1983.
14. P. Erhlich, C. Sagan, D. Kennedy and W. Roberts (eds.), *The Cold and the Dark: the World after Nuclear War*, Sidgwick & Jackson, London, 1984, p. xv.
15. Editorial, see note 8.
16. C. Sagan, 'Nuclear war and climatic catastrophe: some policy implications', *Foreign Affairs*, Winter 1983–84, pp. 257–92.
17. P. Ehrlich et al., see note 14.
18. P. Ehrlich, private communication.
19. One of the many examples of false or misleading reports that the nuclear winter hypothesis had been invalidated is: R. Seitz, *Wall Street Journal*, 5 November 1986. (See letter to *Wall Street Journal*, 24 November 1986, for a devastating reply by the TTAPS group.) Seitz's report was uncritically used as the basis for an inaccurate report in the *Observer*, 30 November 1986. On the other hand, there are also many examples of media reports presenting a nuclear winter as if it were a proven fact.
20. John Maddox, editor of *Nature* was, for example, overhasty in casting doubt on the nuclear winter concept in his editorial of 12 January 1984, in which he seems to overlook the significance of smoke and assume the TTAPS results were based only on the effects of dust. Two distinguished nuclear winter researchers, S. Thompson and S. Schneider, based their article, 'Nuclear winter reappraised', *Foreign Affairs*, Summer 1986, pp. 981–1005, on new, unrefereed computer results they had produced, which indicated that average temperature drops could be as little as 5–10°C. On the basis of this research, the *Foreign Affairs* article sought to

rethink the strategic implications of nuclear winter (or nuclear autumn as they chose to rename it). It was soon pointed out, and acknowledged by Schneider and Thompson, that their calculations depended on a number of unrealistic assumptions about smoke, which were bound to produce milder effects but which no-one believed to be valid.

21. For example, C. Kearny, *Practical Civil Defence*, January/February 1984.
22. National Research Council, see note 7.
23. Royal Society of Canada, *Nuclear Winter and Associated Effects* Report of the Committee on the Environmental Consequences of Nuclear War, Ottawa, 31 January 1985.
24. SCOPE, see note 7; vol. I, A. B. Pittock *et al.*, 'Physical and atmospheric effects'; vol. II, M. A. Harwell and T. Hutchinson, 'Ecological and agricultural effects'.
25. SCOPE, vol. II, see note 7.
26. O. Greene, I. Percival and I. Ridge, *Nuclear Winter, the Evidence and the Risk*, Polity Press, 1985, chapter 2.
27. A. Rip, 'Experts in public arenas', in *Regulating Industrial Risks*, H. Otway and M. Peltu (eds.), Butterworths, London, 1985, pp. 95–110.
28. Meehan, see note 6.
29. P. Ehrlich *et al.*, see note 14, p. 168.
30. The NRC report focused entirely on the potential effects of the nuclear war on the atmosphere and climate. The SCOPE committee had a wider brief, to examine also the potential consequences of a nuclear winter. But, after some internal debate, SCOPE decided against including any extended analysis of the impact on social or psychological processes and structures in its report. (Private communication.)
31. National Research Council, see note 7, p. 2.
32. Ibid.
33. Ibid.
34. L. Dotto, *Planet Earth in Jeopardy: Environmental Consequences of Nuclear War*, John Wiley, 1986.
35. SCOPE Press release, 12 September 1985, Washington, DC, p. 2.
36. *Nature*, vol. 317, 19 September 1985. F. Warner, quoted in T. Beardsley 'Mechanics of SCOPE Report', *Nature*, Vol. 317, 19 September 1985, p. 192.
37. Royal Society of Canada, see note 23, p. 1.
38. Ibid., p. 5.
39. Ibid., p. 7.
40. Ibid., p. 7.
41. The New Zealand Ecological Society, for example, felt able to produce a strongly worded analysis of the possible environmental consequences of nuclear war on New Zealand. Its report, *The Environmental Consequences to New Zealand of Nuclear War in the Northern Hemisphere* (Wellington, New Zealand, December 1984), was explicitly seen by its authors as a statement of concern aimed at alerting New Zealanders to the risks. A number of scientific institutions in New Zealand, including the Royal Society of New Zealand, the Ecological Society and the New Zealand Meteorological Service, began a study on the societal effects on New Zealand of a nuclear winter in October 1986 (published in August 1987). This report extended into economic and policy issues (W. Green, 'The New Zealand impact study', paper presented to SCOPE workshop, Bangkok, 9–12 February 1987). In contrast, the British Ecological Society decided against any society report or investigation into the nuclear winter issue (though it did sponsor a short scientific conference on 15 March 1985.)
42. National Academy of Sciences and National Academy of Engineering symposium

on: 'nuclear winter: current assessment and implications, 26–27 March 1985, Washington, DC.
43. Sagan, see note 16.
44. A concept first introduced by H. Kahn in *On Thermonuclear War*, Oxford University Press, London, 1960; and made famous by the film *Dr Strangelove*.
45. For example, C. Gray, 'The nuclear winter thesis and US strategic policy', *Washington Quarterly*, Summer 1985, pp. 85–96; J. Nye, 'Nuclear winter and policy choices', *Survival*, March/April 1986, vol. 28, no. 2, pp. 119–27; J. Gertler, *Some Policy Implications of Nuclear Winter*, RAND Report P-7045, January 1985; P. de Leon, *Rethinking Nuclear Strategy: Policy Implications of a Nuclear Winter*, Columbia University Press, January 1986; K. Payne, *Strategic Defense: 'Star Wars' in Perspective*, Hamilton Press, London, 1986, chapter 7; M. May, 'Nuclear winter: strategic significance', *Issues in Science and Technology*, Winter 1985, pp. 118–20; G. Rathjens and R. Seigel, 'Nuclear winter: strategic significance', *Issues in Science and Technology*, Winter 1985, pp. 123–7; Lt. Col. D. Drew *et al.*, *Nuclear Winter and National Security: Implications for Future Policy*, Air University Press, Maxwell Air Force Base, Alabama, July 1986.
46. Payne, see note 45.
47. R. Pipes, *US-Soviet Relations in the Era of Detente*, Westview Press, Boulder, Colorado, 1981.
48. C. Kearny, 'Seven reasons why the 'Nuclear winter' capaign is unlikely to result in reduced Soviet nuclear deployments', paper circulated at the National Academy of Sciences Symposium, March 1985, see note 42.
49. Ibid.
50. See, for instance, C. Gray, J. Gertler and K. Payne in note 45. See also F. Hoeber and R. Squire,. 'The nuclear winter hypothesis: some policy implications', *Strategic Review*, Summer 1985; D. Williamson, statement to the Subcommittee on Natural Resources and Environment of the US Congressional Committee on Science and Technology, 14 March 1985; and many other statements to congressional hearings; Col. R. Macdonald, 'point paper' presented to the forum on nuclear winter: strategic and diplomatic implications, Virginia Polytechnic Institute, 6 March 1986.
51. For example, J. Gertler, see note 45.
52. See, for example, E. Teller, *Nature*, 1984.
53. D. Ball, *Targeting for Strategic Deterrence*, Adelphi Paper no. 185, International Institute for Strategic Studies, London, 1983; D. Ball, 'U.S strategic forces: how would they be used?', *International Security*, vol. 7, no. 3, 1982–83, pp. 31–60.
54. The problem being that many prime military targets, such as command and control centres, submarine bases and airfields, are located in or near urban areas: see, for instance, D. Ball, *Can Nuclear War be Controlled?*, Adelphi Paper n. 167, International Institute for Strategic Studies, London, 1981. See also J. Nye, note 45, p. 122.
55. P. de Leon: see note 45, for example, reviews both options while J. Nye and many others prefer the 'no-cities' approach.
56. For example, J. Gertler, D. Drew *et al.*, and P. de Leon, see note 45.
57. T. Postol, 'Strategic confusion—with or without nuclear winter', *Bulletin of the Atomic Scientists*, February 1985, pp. 14–7; D. Drew *et al.*, see note 45.
58. J. Gertler, see note 45.
59. For example, K. Payne and C. Gray, see note 45; C. Weinberger, *The Potential Effects of Nuclear War on the Climate*, US Department of Defense report, March 1985, p. 13.

60. See, for example, C. Sagan, note 16; O. Greene *et al.*, note 26; M. Pentz, 'Policy implications of the nuclear winter hypothesis', paper to conference on the risks of nuclear war, Cardiff, 1984.
61. M. May, J. Nye, and G. Rathjens and R. Seigel, see note 45.
62. Nye, see note 45.
63. Ibid.
64. For example, D. Drew *et al.* and P. de Leon, see note 45.
65. J. Nye, paper presented to NAS symposium on nuclear winter, 27 March 1985, see note 41; A. Gore Jnr, 'Nuclear winter: policy implications', *Issues in Science and Technology*, 1985, pp. 120–3.
66. D. Horowitz and R. Leiber, 'Nuclear winter and the future of deterrence', *Washington Quarterly*, Summer 1985, pp. 59–70; D. Drew *et al.*, and J. Gertler, see note 45.
67. J. Nye, see note 65.
68. R. F. Bacon, memorandum for the chairman, Military Liaison Committee (declassified), 26 July 1983 (reproduced in *The Climatic, Biological and Strategic Effects of Nuclear War*, Hearings before the Subcommittee on Natural Resources, Agricultural Research and Environment, US Congress, Washington, DC, 1985, p. 230–1.
69. Captain L. F. Brooks, Memo. 396, Strategic and Theatre Nuclear Warfare Division, US Navy (declassified), 7 November 1983 (reproduced in *The Climatic, Biological and Strategic Effects of Nuclear War*, see note 68, pp. 222–3).
70. Vice Admiral, J. A. Lyons, Memorandum for the Chief of Naval Operations, 19 November 1983 (reproduced in *The Climatic, Biological and Strategic Effects of Nuclear War*, see note 68, pp. 224–5).
71. Joint Economic Committee, *The Consequences of Nuclear War*, US Congress, 11–12 July 1981.
72. *The Climatic, Biological and Strategic Consequences of Nuclear War*, see note 68.
73. Ibid., 14 March, 1985.
74. Committee on Armed Services, *Nuclear Winter and its Implications*, US Senate, 2–3 October 1985.
75. C. Weinberger, see note 59.
76. Tuft IV (Deputy Secretary of Defense), *Technical Issues Update*, Report to the US Congress, US Department of Defence, Washington DC, March 1986, 5 pp.
77. General Accounting Office, *Nuclear Winter: Uncertainties Surround the Long-Term Effects of Nuclear War*, US General Accounting Office, Report to Congress GAO/NSIAD-86-62, Washington DC, March 1986.
78. See note 74, stenographic transcripts for the first and the second session (denoted I and II in following notes).
79. Ibid., I, p. 5.
80. Ibid., II, p. 4.
81. Ibid., I, p. 74.
82. Ibid., I, p. 5.
83. Ibid., II, p. 47.
84. Ibid., II, pp. 62–3.
85. Ibid., I, p. 8.
86. Ibid., II, p. 24.
87. Ibid., I, p. 39.
88. Ibid., I, p. 45.
89. Ibid., I, p. 27.
90. Ibid., II, p. 41.

91. D. Nelkin, 'Introduction: analysing risk', in *The Language of Risk: Conflicting Perspectives on Occupational Health*, D. Nelkin (ed.), Sage Publications, Beverly Hills, California, 1985, pp. 11–24.
92. See note 74; see, for instance, I, p. 11; I, p. 13; I, p. 60; I, p. 67; I, p. 85.
93. Ibid., I, p. 60.
94. Ibid., I, p. 60.
95. Ibid., I, p. 69.
96. Ibid., I, p. 85.
97. Ibid., I, p. 88
98. Ibid., I, p. 89.
99. Ibid., II, p. 25.
100. Ibid., II, pp. 47–48.
101. Ibid., II, pp. 59–60.
102. Ibid., I, p. 86.
103. Ibid., I, p. 73.
104. Ibid., I, p. 83.
105. Ibid., I, p. 84.
106. Ibid., I, p. 88.
107. Ibid., II, p. 16.
108. Ehrlich *et al.*, see note 14, pp. xviii–xx.
109. D. Healey, *Guardian*, 1 November 1983.
110. For example, correspondence in *Sanity*, July, August and October 1984 following publication of the article by O. Greene, 'The nuclear winter', *Sanity*, June 1984, pp. 16–27; G. Rathjens and R. Seigel, see note 45.
111. C. Meredith, O. Greene and M. Pentz, *Nuclear Winter: a New Dimension for the Nuclear Debate*, SANA, 1984, 43 pp.; Nuclear Winter Scientific Workshop, Oxford University, 14 January 1984; Nuclear winter conference, Sheffield, 9 June 1984; *A Change in the Weather*, SANA video cassette, presented by M. Pentz and I. Ridge, 1984.
112. For instance, Mrs Fenner, written answer to Mr Home Robertson, *Commons Hansard* 29 October 1984, col. 859–60.
113. 'Critique of nuclear winter research', (typescript), 2 November 1984, enclosed in 'TTAPS national briefing on the nuclear winter', Home Office circular to emergency planning officers, 2 November 1984. A version of Kearny's critique was published as 'The "other" nuclear freeze: will a full-scale or limited nuclear war initiate another ice age?' *Practical Civil Defence*, January/February, 1984, pp. 14–8, and also in other civil defence journals.
114. The Scottish Office also declined a request that its scientists concerned with civil defence questions be allowed to attend these briefings, as justified by Mr Ancram in a written answer to Mr Dalyell, *Commons Hansard*, 13 November 1984, col. 207.
115. C. Sagan, quoted in *World Wide Consequences of Nuclear War*, Nuclear Freeze Foundation, 1984, p. 75.
116. C. Weinberger, see note 59, pp. 11, 12.
117. S. Shenfield, 'Nuclear winter and the USSR', *Millenium*, vol. 15, no. 2, 1986, pp. 197–208.
118. 'Uchenye predosteregayet', *Pravda*, 23 November 1983, referred to in S. Shenfield, see note 117.
119. Shenfield, see note 117, footnote 49.
120. B. Gontarev (ed.), *The Night After: Climatic and Biological Consequences of Nuclear War*, Mir Publishers, Moscow, 1985.
121. Shenfield, see note 117.
122. S. Shenfield, *The Nuclear Predicament: Explorations in Soviet Ideology*, Royal Institute of International Affairs, 1987, p. 15.

123. D. Holloway, *The Soviet Union and the Arms Race*, Yale University Press, 1983.
124. Ibid.; Shenfield, see note 117.
125. A. Yakovlev, 'Confrontation is anomaly in international relations', *Twentieth Century and Peace*, no. 2, 1985, p. 24.
126. Quoted by Shenfield, see note 117, p. 11.
127. Gorbachev speech to the Moscow Peace Forum, February 1987.
128. Nye, see note 65.
129. UN Resolution 39/148F, proposed by India, Mexico, Pakistan, Sweden, Uruguay and Yugoslavia.
130. R. Alfonsin, R. Gandhi, M. de la Madrid, J. Nyerere, O. Palme and A. Papandreou, Delhi Declaration, 28 January 1985.
131. Nye, see note 65.
132. W. Green, T. Cairns and J. Wright, *New Zealand after Nuclear War*, New Zealand Planning Council, Wellington, 1987. See also note 41.
133. Green, private communication.
134. Ministry for the Environment, New Zealand, *What Will Happen to New Zealand if There Is a Nuclear War?*, 1987.
135. Postol, see note 57.
136. See, for example, the statements of senior officials in the incoming Reagan administration as quoted in R. Scheer, *With Enough Shovels: Reagan, Bush and Nuclear War*, Random House, New York, 1982.

2

Rendering Nuclear Weapons Impotent and Obsolete: the Origins of a Pipedream

DAVID CARLTON

In the presidential campaign of 1976 President Gerald Ford in a crucial televised debate with Jimmy Carter indicated, in reply to a question, that he considered Poland not to be a member of the Warsaw Pact. Of course, politicians being what they are, most members of his administration would have had no moral objection to echoing their chief if he had asked them to do so. But, as luck or ill-luck had it, they were on this occasion spared from having to make idiots of themselves in public. For Ford's blunder was so crass and so easily demonstrable that he had no rational alternative but to admit his error. On 23 March 1983 President Ronald Reagan, in a celebrated television address, also revealed his fallibility when he said the following:

> What if free people could live secure in the knowledge that their security did not rest upon the threat of instant United States retaliation to deter a Soviet attack, that we could intercept and destroy strategic ballistic missiles before they reached our own soil and that of all allies? . . . I call upon the scientific community in our country, those who gave us nuclear weapons, to turn their great talents now to the cause of mankind and world peace, to give us the means of rendering these nuclear weapons impotent and obsolete.[1]

In Reagan's case, in contrast to that of Ford, however, his blunder was only self-evident to experts rather than to the American nation as a whole. Accordingly all members of his administration were predictably required to endorse or at any rate refrain from contradicting the line of their chief. All appear to have been prepared to comply—with one notable and honourable exception. The exception was Richard DeLauer, then Under Secretary of Defense for Engineering. In May 1983 he pointed out that the proposed defensive system could not be achieved in the absence of Soviet co-operation. 'With uncontrolled proliferation' of Soviet warheads, he bluntly stated, 'no defensive system will work'.² And in the summer of 1983 he said 'There's no way an enemy can't overwhelm your defenses if he wants to badly enough. It makes a lot of difference in what we do if we have to defend against 1,000 RVs [re-entry vehicles] or 10,000'.³ He soon found himself out of office and has subsequently underlined in public his astonishment at the content of Reagan's speech.

The fact is that no significant number of independent scientists, whether for or against deploying limited strategic defences, have dissented from DeLauer's judgement. Many have also argued that even if the Soviets exercised restraint in producing further nuclear warheads, no American defensive deployments could ever guarantee 100 per cent protection against intercontinental ballistic missiles (ICBMs). Moreover, there are other means of delivering nuclear weapons which Reagan's Strategic Defense Initiative (SDI) speech effectively ignored. Air-breathing cruise missiles present an obvious problem; and even if these could be neutralised by a non-space-based defensive system there would still be the 'suitcase bomb' to worry about: not only the Soviet Union but many other states might in the future succeed in assembling a nuclear weapon in the United States simply by smuggling in the components. How, then, can nuclear weapons possibly be rendered impotent and obsolete?

In the course of a visit to Washington in 1986 I asked an expert close to the Reagan administration to comment on an unattributable basis on these arguments. He made no attempt to conceal his own belief that Reagan had indeed offered his nation a utopian vision that was simply unattainable. But he added that a 'lawyer's argument' might be made that the President had not really meant to hold out the hope that the United States could by its own unaided efforts achieve 100 per cent protection against nuclear attack. What he might have had in mind, according to this argument, was merely to create powerful if imperfect defences which could then be rendered perfect as a result of a subsequent nuclear disarmament agreement negotiated with the Soviet Union (and maybe other nuclear weapon states).⁴ I agreed that the wording of his speech (on a strained interpretation) might just be held to be compatible with this version. But my rejoinder was to point out that the President had not subsequently explained in clear language that this was what he had meant and he had had numerous opportunities to do so. Moreover, Caspar Weinberger, his Secretary of Defense, stated on 27 March 1983:

> The defensive systems the President is talking about are not designed to be partial. What we want to try to get is a system which will develop a defense that is thoroughly reliable and total. . . . I don't see any reason why that can't be done.⁵

Confronted with this quotation my well-connected source could only offer the comment that the President might not have been best served by his over zealous Defense Secretary! Perhaps not. But the fact remains that Reagan has not repudiated Weinberger's words; and, whether dishonestly or not, has permitted an unattainable utopian vision to confuse the only serious debate, namely the one about whether *limited* strategic defences are desirable.

In the remainder of this paper it is proposed to ask why President Reagan saw fit to introduce this utopian note into his speech of 23 March 1983. Did he get the idea from any pro-ABM advocates or pressure group? Or did he bring the idea into the White House as a result of some earlier personal experience as Governor of California or as a presidential candidate? Above all there is the question of the scientific and technological advice he received as to the feasibility of rendering nuclear weapons impotent and obsolete. Did he seek such advice? If so, what was its nature? Did he ignore any such advice?

It remains to be acknowledged, by way of introduction, that any historian considering such a recent development faces obvious problems. But such is the open nature of American society and government that it is already possible to offer at least tentative answers to some of the foregoing questions.

Reagan as governor and presidential candidate

Our first task is to try to decide whether Reagan brought his utopian vision with him into the White House in 1981 as a result of some earlier personal experience. Those who speculate that this could have been the case point to two episodes that conceivably could have been important.

The first occurred in 1966 shortly after he became Governor of California. He accepted an invitation to visit the weapons laboratory at Livermore, California, whose chief was the hard-line nuclear physicist Dr Edward Teller. According to Teller's own account:

> He [Reagan] listened carefully; not to a highly technical presentation, but to one that must have contained a host of completely novel ideas. He asked maybe ten or twelve questions which clearly showed that he followed—that he comprehended. Indeed, he was the only governor who ever visited our laboratory.[6]

As is well known, Teller is opposed to mutual assured destruction (MAD) and has long favoured research on various defensive capabilities. Could it be, then, that he became a guru for Reagan and that herein lies an explanation for the speech of 23 March 1983? It is indeed possible that Teller had much to do with Reagan's acceptance of the alleged bankruptcy of a strategy based on total mutual vulnerability between the superpowers. But this does not explain the utopian reference with which we are here concerned. For Teller appears never to have had any illusions of that kind. And in 1985 he categorically stated in an interview with the BBC: 'They say a complete defence is impossible. Of course it is impossible.'[7]

A second episode to which speculative attention has been paid concerns Reagan's visit to NORAD while he was campaigning to become Republican presidential candidate. He was much struck by the implication of total American vulnerability to Soviet ICBMs, which he appeared to have fully grasped for the first time. He is said to have drawn anguished analogies with gunfighters in Western films.[8]

He unburdened himself to a journalist, Robert Sheer of the *Los Angeles Times*, while on the campaigning trail in 1980:

> NORAD is an amazing place—that's out in Colorado, you know, under the mountain there. They actually are tracking several thousand objects in space, meaning satellites of ours and everyone else's, even down to the point that they are tracking a glove lost by an astronaut that is still circling the earth up there. I think the thing that struck me was the irony that here, with this great technology of ours, we can do all of this yet we cannot stop any of the weapons that are coming at us. I don't think there's been a time in history when there wasn't a defense against some kind of thrust, even back in the old-fashioned days when we had coast artillery that would stop invading ships if they came.[9]

But Sheer's account of the subsequent conversation, already in print *before* the SDI speech, has Reagan musing on the need for an American civil defence programme of the kind supposedly already possessed by the Soviet Union.[10] He did not, however, speak of the possibility of creating leak-proof shields over the two superpowers. It seems reasonable to conclude, therefore, that Reagan did *not* bring his utopian plan into the White House in 1981 as a result of the trauma caused by his visit to NORAD, even though the concerns about American vulnerability that led to it were clearly already present.

One other aspect of Reagan's pre-presidential experience merits attention. It is that in the matter of national security issues he had a longstanding guru noted for his strong opinions but not generally greatly respected in the strategic studies community. This was Laurence Beilenson, a Californian lawyer who had advised the Screen Actors Guild when Reagan was its anti-Communist President in the immediate post-war years. He is the author of a book entitled *The Treaty Trap* which Reagan commended to West Point cadets in May 1981. Its general thesis is that arms control agreements with the Soviets are futile. But while Beilenson favoured abandonment by the United States of the ABM Treaty of 1972, he does not appear to have contended that nuclear weapons can simply be rendered impotent as threats to populations by defensive technology. His outlook is summarised in an article he co-authored in January 1982 for the *New York Times Magazine*:

[A new nuclear strategy] will call for developing offensive weapons in great quantities, and exploiting mobility, concealment and dispersion to make them invulnerable to any conceivable nuclear strike. It will call for removing the shackles placed by the strategic arms treaty of 1972 on the development of anti-missile weapons capable of protecting our population and economy. It will call for restoring our emaciated air defense against bombers so as to meet an impending Soviet threat in that category of weapons. And it will require a program, following the Swiss example, for providing our cities with civil defense shelters, which can save enormous numbers of American lives. . . .[11]

The very fact that Beilenson envisaged the need for civil defence shelters and increased United States offensive forces suggests that he did not foresee at this stage the creation of leak-proof shields over either of the superpowers. So Beilenson, however influential he may once have been, does not appear to have been the Svengali behind the utopian vision of the President's speech of March 1983. In short, the utopianism was almost certainly a product of developments in his thinking *after* he entered the White House.

The role of American pressure groups

When Reagan entered the White House several prominent conservative pressure groups naturally assumed that their years in the wilderness were over and that they would soon see their cherished schemes translated into action. They had in fact more reason than similar groups in the past had had to travel hopefully. For Reagan and his immediate entourage appear to have been the most ideologically motivated team to move into the presidency in modern times.

In practice, however, the zealots have had their disappointments, not least in matters relating to defence and foreign policy. For entrenched bureaucratic interests in the Department of Defense, at the State Department and even at the Central Intelligence Agency have ensured a fair amount of continuity with the Carter years. Hence one has heard calls to 'allow Reagan to be Reagan' and jibes to the effect that Alexander Haig and George Schultz have presided over the People's Republic of Foggy Bottom. Yet the zealots have not found the experience of the Reagan presidency entirely negative. In short, *some* policies have changed since 1980. Moreover, large numbers of the personnel of the various pressure groups have been given positions of power and influence in or on the edges of the administration and in some cases they themselves have played a part in accepting policies they disapproved of when in opposition. In the light of this general setting, let us now consider how the pressure groups and their leading personalities have conducted themselves in the matter of Reagan's utopian vision for rendering nuclear weapons impotent and obsolete.

Did any of them sponsor the idea before or after Reagan came to power and how did they in the event react to his SDI speech?

The most important conservative pressure group concerned exclusively with national security issues in the pre-Reagan years was the Committee on the Present Danger founded in 1976. Its acknowledged leader was Paul Nitze, an outstanding public servant since the days of President Harry S. Truman (to whom he had presented the famous NSC 68 analysis urging an active policy for the containment of international communism). Others involved included Eugene Rostow, Richard Perle, Edward Rowny, Fred C. Ikle, Richard Allen, Richard Pipes, William Casey and Jeane Kirkpatrick (all to become members at varying levels of seniority in the Reagan administration when it was eventually formed). In opposition they undoubtedly helped to create a climate favouring robust approaches to national security policy; and in office they were usually to be found on the 'hawkish' side of the numerous debates within the administration. What cannot be claimed, however, is that the Committee or any of its prominent personalities canvassed in opposition the idea of trying to render nuclear weapons 'impotent and obsolete'. Essentially they seem to have favoured a major American nuclear arms build-up to close the alleged window of vulnerability, but with no particular collective bias in favour of defensive as distinct from offensive weapons. When in office they appear to have accepted a lead from Reagan in the matter of the SDI rather than to have been responsible for pushing him in that direction. Moreover, most have been inclined in supporting the SDI to emphasise its medium-term implications rather than the long-term, utopian goal with which we are here concerned.

A second conservative grouping was the National Institute for Public Policy based at Fairfax, Virginia. Its leading figures were (and are) Colin S. Gray and Keith Payne. This institute has long been associated with hostility to the Anti-Ballistic Missile Treaty of 1972 and undoubtedly welcomed Reagan's SDI speech. Their support for the President, however, also seems to have been based on the entirely reasonable belief that the effect of his initiative would not be the establishment of leak-proof nuclear shields over the two superpowers but would rather be to create defences for hard-point targets, that is the American ICBMs, and maybe also provide some *limited* protection for the general population.

Gray and Payne, in their various writings before and after March 1983, have revealed great and consistent concern for maintaining and enhancing 'extended deterrence', that is, the American guarantee of the security of Western Europe. To them the threat of a possible first use of nuclear weapons by the United States is a *sine qua non* for the maintenance of the credibility of such guarantees; and they even foresee circumstances in which the United States could prevail in a nuclear war. Accordingly, for example, Gray was a severe critic of the proposal for a NATO no-first-use declaration advanced by McGeorge Bundy, George F. Kennan, Robert S. McNamara and Gerard Smith.[12] Gray wrote of this idea in 1982: 'Conflict in Europe must always be conducted in the shadow of nuclear weapons, no matter what NATO's declaratory policy may be.'[13] We may surmise, therefore, that Gray believes it to be neither possible nor desirable that Reagan's utopian vision should become a reality.

Payne, too, is in no way to be seen as an advocate of rendering nuclear weapons impotent and obsolete. In 1983, for example, he wrote an article urging revision of the ABM Treaty. But his aim was to see defences that would make ICBMs less vulnerable and afford a degree of population protection. He specifically added that 'this is not to argue that an impenetrable area defense system is now, or ever will be feasible'.[14]

The National Institute for Public Policy is thus clearly not to be held responsible for in any way contributing to Reagan's move towards utopianism. Indeed, we may conclude that this particular aspect of his SDI speech runs strongly counter to the general strategic approach of that institute.

The National Institute for Public Policy is not noted, however, for having had any single-minded concern with promoting space-based nuclear defences. The organisation which has that reputation is High Frontier. So it may be particularly appropriate to examine its origins and outlook. It is in reality the creation of one possibly obsessive individual, namely Daniel O. Graham, a retired Lieutenant-General and former head of the Defense Intelligence Agency. He has long believed that the Soviets are determined to take the arms race into space and believes that the United States can and must do the same. He is, moreover, extremely optimistic about the technological feasibility of creating quite impressive space-based defences in the immediate future.

Graham launched his personal crusade for space defences while the Republicans were still in opposition and found some sympathetic listeners at the Heritage Foundation, the prominent Washington-based conservative think tank. Soon after Reagan's inauguration, a panel was established, with Heritage Foundation funding, to attempt to draw up precise recommendations for the new administration. Much to Graham's chagrin, however, the panel came to be dominated by supporters of Edward Teller. In particular, Karl R. Bendetsen emerged as its leading figure. Another key personality was Lowell J. Wood, a close associate of Teller and a principal figure in the Hertz Foundation. The result was a policy split which led to Graham's forming his own organisation, namely the Washington-based High Frontier.

The split, however, was not about whether perfect defences could be established. Rather the dispute was about the type of space-based defence that should be recommended to the administration and the likely timescale involved. Graham thought rapid moves could be made in the direction of selecting 'off-the-shelf' technology. Teller's supporters favoured an emphasis on longer-term research, believing that Graham's short cut would be so technologically flawed as to be potentially useless for any serious defensive purpose.

Though Graham had a meeting with Reagan and subsequently received a friendly letter from him in June 1983 commending him for 'the important work that you and your colleagues have done to prepare the way for a more secure America',[15] it would appear that he had much less influence than the Heritage-based majority who supported Teller. Teller had a number of meetings with the President, which may indeed have been decisive in shaping the longer-term emphasis on research that has been the main practical outcome so far of the SDI speech.

Graham, however, continues to fight his corner and also claims to be a strong supporter of SDI in principle. For our purposes, however, it is important to note that neither before nor after Reagan's SDI speech did Graham raise hopes in the mind of the President or anyone else that leak-proof shields could be established. In 1983 he published *We Must Defend America and Put and End to MADness* in which these words appear.

> If effective strategic defense is defined as an impenetrable shield through which no nuclear weapons could penetrate, there will be no defense. Obviously, no defense in history has ever been perfect. . . . High Frontier rejected absolutist demands on strategic defense and looked instead for defensive systems and programs which could add significantly to the U.S. deterrent to nuclear war.[16]

This work was, however, written before the President's SDI speech.[17] How, then, did Graham react to its utopian element? With admirable consistency he has not overnight abandoned his former approach. Instead, he has been able to support Reagan by seeing fit to suppose that 100 per cent defense was not envisaged in the SDI speech. For example, on 16 July 1986 he wrote as follows to a number of United States legislators:

> The notion that only 100% perfect defenses can be considered protection for population is a false premise cultivated by the anti-SDI lobby and trumpeted in the media. It is a polemical device for undermining broad popular and political support for the defense systems. It allows anti-SDI spokesmen to support continuation of adherence to the MAD doctrine without saying so.
>
> Another false impression often heard of late is that the President originally called for a 'leakproof umbrella', but has changed his mind and now wants merely to protect missiles. Purveyors of this idea would have us believe that when General [James] Abrahamson, the SDIO Director, states that we cannot achieve a perfect defense, or when Richard Perle says that we will start with defenses that reinforce

deterrence more than protect population, they are at odds with Secretary Weinberger who stresses the goal of defending the people.

This is all quite erroneous. President Reagan never called for perfection in strategic defenses, and neither has anyone else who is *for* SDI. Abrahamson, Perle, Weinberger all know, and have often stated, that the defenses will progress in stages, and that early stages will be more effective in the deterrent role than in the population role, but that the end goal of SDI is effective population protection—not perfect, but effective.[18]

Readers will realise that the present writer does not think that either Reagan or Weinberger has in fact taken the line attributed to him by Graham. But for our present purposes what is important is that Graham and High Frontier cannot possibly be held even partially responsible for the call to render nuclear weapons impotent and obsolete.

The majority group operating under Heritage Foundation auspices seems likewise to have been uninvolved in utopianism. For as we have already seen, their principal inspirer, Teller, flatly stated to the BBC in 1985 that 'complete defence is impossible'.[19] In short, they have differed from Graham but not on this aspect of the matter.

As for the Heritage Foundation itself, it was naturally in the vanguard of those anxious to see the abandonment of the constraints on defence enshrined in the ABM Treaty. But nothing in many pre-1983 Heritage Foundation publications dealing with this theme gave any indication that utopian aspirations were involved. Since 1983 the utopian issue has usually been glossed over, doubtless to avoid giving the appearance of their differing from Reagan.

As well as these pressure groups there were many individual strategic analysts arguing in the late 1970s and early 1980s in favour of a move towards nuclear defences in one form or another. The present writer has accordingly explored this literature on a quite extensive scale. But no article or book advancing the utopian goal embraced by Reagan has come to light. Overwhelmingly the emphasis was on matching Soviet efforts; on protecting ICBMs; or on acquiring *limited* population defences for a variety of reasons, including the possibility of countering accidental, catalytic or Nth power strikes.[20] It seems fair to conclude, therefore, that in the quest for the origins of Reagan's utopian vision we must look to the inner sanctums of his administration.

The Joint Chiefs, the National Security Council, the Chief Scientific Adviser and the President

Any attempt to explain what happened in the White House in the period before March 1983 is bound to be tentative in character. For none of those centrally involved has so far written memoirs or published diaries. Furthermore, two of the principals, the President himself and Robert McFarlane, have given only sketchy press briefings on the matter. So we are very much dependent on one major source, namely the then Presidential Chief Scientific Adviser George A. Keyworth II; and even Keyworth, in interviews with well known American newspapers, has understandably shown an inclination to be less than totally candid.

I have not myself had an opportunity to meet Keyworth. But I was given a most helpful hint by one who was in a position to know Keyworth's side of the story. He pointed me in the direction of an article in a less well known American newspaper, the *Philadelphia Inquirer*, and he gave me to understand that the revelations in it could be taken to be reasonably accurate. The journalist responsible for the article, Frank Greve, had had an interview with Keyworth, who had apparently seen fit to be unusually forthcoming.[21] What follows is based partly on Greve's article, though some details have been independently confirmed by various 'insiders'.

The first aspect to consider is the role of the Joint Chiefs of Staff. Reagan's speech of 23 March 1983 contained this passage:

> Wouldn't it be better to save lives rather than avenge them? Are we not capable of demonstrating our peaceful intentions by applying all our abilities and all our ingenuity to achieving a truly lasting stability? I think we are. Indeed we must.
>
> After careful consideration with my advisers, including the Joint Chiefs of Staff, I believe there is a way. Let me share with you a vision of the future which offers hope.[22]

That the Joint Chiefs were in principle interested in intensifying the exploration of the possibilities for strategic defences is clear. Like many experts, they cannot have been happy with the way in which the strategic stability expected to result from the ABM Treaty of 1972 had not materialised as a result of the unexpected emergence of the threat posed to the American ICBM force by the vast array of Soviet MIRVed missiles. They also could have been in no doubt that the Soviets had long been intensively involved in researching into defensive technology. But it is not clear, as Reagan implied, that they were properly consulted about his famous speech or that it reflected their own thinking. Greve wrote on the basis of his interviewing:

> 'We recognised that SDI was not a panacea' against nuclear threats of all kinds, said retired Army General John W. Vessey Jr, then chairman of the Joint Chiefs. In fact, the Joint Chiefs had urged the President to consider strategic defense, but they had no specific plan in mind, simply considering it a concept worth careful study.[23]

As for presentation, they might well have preferred to 'sell' the matter of spending more on defensive research in the traditional 'action–reaction' way, that is to have called it not a strategic defense initiative but a strategic defense response—given what was known of the intensifying Soviet efforts. Moreover, American research on defensive technology had actually been increased during Carter's time; so a further increase could have been presented more in terms of continuity than of strategic revolution. But the fact is that Reagan had his own ideas on presentation and in no way could the Chiefs be said to have been responsible for this. In particular, the utopian aspect, which is our main concern in the present article, was not something which the Chiefs, with the possible exception of Admiral James Watkins, who had had some prior meetings with McFarlane, appear to have been asked to give any serious and detailed consideration. The actual text of the SDI speech, according to Greve, was sent to them a mere two days before delivery.[24] It thus need not surprise us that the Chiefs have not subsequently revealed any marked faith in the technological feasibility or even desirability of rendering nuclear weapons impotent and obsolete.

If the Chiefs were presented with something like a *fait accompli* the same went for other key personnel. According to Greve, Secretary of State Schultz received two days notice; while Fred Iklé and Richard DeLauer at the Pentagon received a mere nine hours.[25]

What, then, about the role of Chief Scientific Adviser Keyworth? He had been recommended to the President by Teller and was genuinely supportive of the idea of exploring strategic defensive possibilities. But he appears to have had serious doubts about the way in which Reagan's SDI speech was planned and launched. According to Greve, he received only five days notice of the President's broad intentions and 'he might have gotten less' according to a National Security Council member, 'had not we asked ourselves, "How can the President go on the tube directing a major high-technology initiative and tell his scientific adviser nothing?"'[26] Greve's account of Keyworth's role continued:

> His immediate reaction to the Star Wars idea, was, 'Give me time. It's big. Give me time.' Keyworth said in a recent interview. 'Most people saw the speech very close to the time of delivery and most—myself included, incidentally—had the same reaction: "My God, let's think about this some more. Let's think about the implications for the allies. Let's think about what the Soviets are going to think. Let's think about what the scientists are going to think. Let's think about the command and control problems."'[27]

Keyworth's pleas for more time for exploration and preparation were, however, rejected. Reagan evidently believed, probably correctly, that a top-down lead was the only way to prevent the project from being stymied by existing vested interests in the great bureaucracies of State and Defense, not to mention NATO. Whether his top-down lead needed to be quite so underprepared is, of course, another matter.

Reagan in fact made the decision on the final contents of his speech with an absolute minimum of expert counselling. Possibly a few

National Security Council (NSC) staff, in particular McFarlane, gave some advice on the possible political impact, though they may have been more interested in seizing the 'moral high ground' from the Freeze Movement and the Roman Catholic bishops than in considering the impact on broad Western security concerns. But on the scientific and technological side almost nothing of a detailed nature seem to have been put to the President in the final and vital drafting stage. True, Reagan at one point asked Keyworth to prepare a text for his proposed speech. But he found the result too bland. The final draft, including the utopian passages, appears, then, to have been his own work, possibly assisted by a handful of NSC cronies who may have proposed a few minor amendments to take minimal account of last-minute reactions from those in the administration hitherto kept in the dark. The competence of NSC staff to be the sole guides to the President on the final form of so important a speech is, to say the least, not self-evident. But this point may need no underlining in the light of the disgracing of the NSC in the Iranian affair of 1986–87.

Keyworth did not resign over the SDI speech, although he did later. For, as a protégé of Teller, he was, as stated, sympathetic to the principle of exploring defensive possibilities. But, like many others with that approach, he clearly wished that the matter had been differently handled.

One detects something of the same ambivalence in the work of the two study groups subsequently established by Reagan to explore some of the detailed implications of his speech. In particular, both groups, headed respectively by James C. Fletcher and Fred S. Hoffman, obviously had difficulty in taking entirely seriously the utopian aspects of the President's speech but loyally sought to gloss over this fact. Presumably most enthusiasts for strategic defences hope that the utopianism will be forgotten when the President leaves office but that the essentials of the SDI research programme will endure. In short, they hope that Reagan's cavalier approach to technological possibilities will prove in the end to have been of no lasting importance.

Conclusion

How important in fact was Reagan's lurch into utopianism? At this point only a tentative answer can be offered. But it does indeed seem possible that after an initial flurry it has largely ceased to matter.

First we may ask whether the utopianism made any difference to American domestic politics? Assuming a prime motive was to seize the 'moral high ground', was it successful in so doing? Certainly the SDI vision had a strong appeal for many Americans. However, the present writer holds the somewhat unfashionable view that no decisive political advantage was gained overall. Most of those Americans who were impressed by the Freeze Movement and/or the Roman Catholic bishops' letter may well remain sympathetic to their ideas. In any case, they are unlikely, in my opinion, to have been converted in any large numbers on the basis of one ill-prepared speech to support for strategic defences. How, then, is one to account for the fading of the Freeze Movement? For the answer one should probably look at the history of similar mass movements in the United States and other Western countries, not least the United Kingdom and West Germany. The fact is that all such movements find it hard to maintain momentum. Vigorous, some may say fashionable, for a couple of years, they then tend to go into periods of hibernation as particular reasons sparking intense activity lose their immediacy. Reagan's SDI vision did allow him to take the political offensive in one of the few areas where he had previously been vulnerable—presenting alternatives to a nuclear arms race—and could therefore have hastened the decline of the Freeze Movement. However, his landslide victory in the 1984 presidential election was, we can now see, heavily overdetermined. His SDI line, in contrast to Walter Mondale's restrained and qualified endorsement of the nuclear freeze idea, would seem to have been at best of only marginal significance. This opinion is bolstered by examining the results of the congressional elections of 1986, when Reagan's attempt to use SDI to boost the chances of Republican candidates was unsurprisingly of no avail in circumstances when a big swing to the Democrats was also overdetermined for reasons largely unconnected with the arms race.

What, then, about the effect of Reagan's utopianism in creating tensions with the West European allies—sometimes seen as the most important result of his speech? To be sure, the initial reaction in NATO was one of incredulity and hostility. Various public attacks, in code or otherwise, were made by Western European leaders and we may guess that behind closed doors fiercer rebukes were issued. Among Western European strategic studies experts, too, strong disapproval was often expressed with vigour and indignation. Consider, for example, the case of the International Institute for Strategic Studies. It is said to have heard somewhat acrimonious exchanges in the course of its annual conference at Avignon, France, in September 1984—something quite unusual in that citadel of Western 'establishment' defence thinking. Furthermore, in its first quarter century of existence few if any papers appeared under its auspices that were so unrestrainedly critical of the leader of the Western world as was that presented at Avignon by Lawrence Freedman, Professor of War Studies at King's College, London. In a memorable passage he delivered this severe verdict:

> President Reagan's speech of March 1983 may have launched a thousand research projects but it did not launch a strategic revolution. He was offering a false prospect of invulnerability, an illusion that he had some bold escape plan from the harsh realities of the nuclear age. This would have quickly been dismissed as the ramblings of a sentimental idealist had he not been President of the United States and had he not backed up his vision with the promise of a technical solution that was soon to be found wanting.[28]

Yet by Reagan's second term most Western European leaders and strategic analysts had calmed down. For the realisation seems gradually to have dawned in Europe that mere words by Reagan could not alter technological realities. There will be no leak-proof shields over the superpowers. And thus 'extended deterrence' is not likely to lose whatever credibility it has; at any rate not on account of the utopianism in the SDI speech. Likewise the ability of the British and the French to deliver 'last resort' nuclear strikes on the Soviet Union is not about to disappear. True, Reagan has more than once caused further alarm in London and Paris by offering with droll consistency to share the US defensive capability with the Soviets. Yet as perfect defences are not what would be shared (if ever so unlikely a deal turned into reality), the two Western European nuclear weapons states, which are both in the process of greatly strengthening their firepower, are probably in no real danger of losing the capacity to inflict unacceptable damage on the Soviet Union. That is not to say that the Western Europeans are not justified in holding strong views about the merits and demerits of developing partial defences and of moving away from unqualified mutual assured destruction. But they are arguments that do not relate to any totally transformed strategic environment in the foreseeable future.

There has, in short, been a gradual recognition throughout the West that utopianism in Reagan's mouth is still utopianism. After all, nearly every Western leader for more than a quarter of a century has paid lip service in company with the Soviets to the ideal of general and complete disarmament; and literally hundreds of diplomats have wasted thousands of hours in posturing at Geneva and New York in support of this unattainable objective.[29] Yet none of this has caused the slightest alarm to supporters of 'extended deterrence' or of Western European independent nuclear capabilities.

Probably the reason that many Western Europeans reacted differently in the case of Reagan's speech is that they sensed that he might actually have meant what he said. But second thoughts seem to have led to the realisation that technological realities are not affected even if he did and does. The way is thus now clear for the only serious debate that has ever been relevant in this area, namely whether *limited* strategic defences are worth pursuing.[30]

Notes

1. The full text of Reagan's speech is conveniently accessible in: Office of Technology Assessment, *Strategic Defenses: Ballistic Missiles Defense Technologies; Anti-Satellite Weapons, Countermeasures and Arms Control: Two Reports by the OTA*, Princeton University Press, Princeton, New Jersey, 1986, Part One, pp. 297–98.
2. Quoted in: Strobe Talbott, *Deadly Gambits: The Reagan Administration and the Stalemate in Nuclear Arms Control*, Vantage Books, New York, 1985 ed., p. 318.
3. Quoted in: Lawrence Freedman, 'The "star wars" debate: the western Alliance and strategic defence', in International Institute for Strategic Studies, *New Technologies and Western Security Policy*, Adelphi Paper, no. 199, London, 1985, p. 49, n. 50.
4. The line of argument, if Reagan actually believes it, is surprisingly close to that of Jonathan Schell, *The Abolition*, Knopf, New York, 1984.
5. Quoted in Freedman, see note 3, p. 37.
6. Quoted in: Michael Charlton, *The Star Wars History: From Deterrence to Defence: The American Strategic Debate*, BBC Publications, London, 1986, p. 95.
7. Ibid., p. 111.
8. Frank Greve, 'Out of the blue: how "star wars" was proposed', *Philadelphia Inquirer*, 17 November 1985.
9. Robert Scheer, *With Enough Shovels: Reagan, Bush and Nuclear War*, Random House, New York, 1982, p. 104.
10. Ibid., pp. 105–7.
11. Laurence W. Beilenson and Samuel T. Cohen, 'A new nuclear strategy', *New York Times Magazine*, 24 January 1982; see note 9, p. 180, n. 26.
12. McGeorge Bundy, George F. Kennan, Robert S. McNamara and Gerard Smith, 'Nuclear weapons and the Atlantic Alliance', *Foreign Affairs*, vol. 60, no. 4, 1981–82, pp. 753–68.
13. Colin S. Gray, 'NATO's nuclear dilemma', *Policy Review*, no. 22, Fall 1982, p. 109.
14. Keith Payne, 'Should the ABM treaty be revised?', *Comparative Strategy*, vol. 4, no. 1, 1983, pp. 1–20, and especially p. 10.
15. Ronald Reagan to Daniel O. Graham, 3 June 1983, quoted in: Daniel O. Graham, *High Frontier*, Tom Docherty Associates, New York, 1983.
16. Daniel O. Graham, *We Must Defend America and Put an End to MADness*, Regnery Gateway, Chicago, 1983, pp. 47–8.
17. Interview with Daniel O. Graham, 17 July 1986.
18. Daniel O. Graham to several American legislators, 16 July 1986. I am most grateful to General Graham for making this letter available to me.
19. See note 7.
20. See, for example, numerous articles in a special issue of the *Washington Quarterly* entitled *ABM revisited: Promise or Peril*. This resulted from the assembling of nine experts at the Center for Strategic and International Studies in Washington. (*Washington Quarterly*, vol. 4, no. 4, 1981).
21. Greve, see note 8.
22. Office of Technology Assessment, see note 1, p. 197.
23. Greve, see note 8.
24. Ibid.
25. Ibid.
26. Ibid.
27. Ibid.
28. Freedman, see note 3, p. 47.
29. See David Carlton, 'International systemic features inhibiting disarmament and

arms control', in: David Carlton and Carlo Schaerf (eds.), *Reassessing Arms Control*, Macmillan, London, 1985, pp. 28–35.
30. I am grateful to the Faculty of Technology at the Open University, of which I was then a member, for awarding me a travel grant that allowed me to carry out research and interviews for this article in Washington in the summer of 1986. I presented an earlier version to the Rome-based International School on Disarmament and Research on Conflicts in August 1986 at its San Miniato conference, and I wish to thank participants for their helpful comments and criticisms. Likewise I am grateful to the editors of the present volume for their suggestions. My thanks are also due to the following experts in the United States who granted me interviews: Daniel O. Graham, Keith Payne, Jack Mendelsohn (the Arms Control Association), Robert L. Schuettinger, and several persons who preferred not to be identified. None bears any responsibility for the conclusions I have reached.

3

The Alchemists of Our Time: the Weapons Scientist as Scapegoat

MARGARET BLUNDEN, OWEN GREENE AND JOHN NAUGHTON

Introduction

In a celebrated address to the American Philosophical Society in Philadelphia in November 1979—since described as 'easily the most important public statement by any British scientist at any time on the subject of nuclear weapons'[1]—Lord Zuckerman reflected upon the motive forces driving the nuclear arms race. In doing so, Zuckerman, former Chief Scientific Adviser to, successively, the Ministry of Defence and the British Government, focused on the process of military research and development, and on the people who do this kind of work in weapons laboratories.

Whereas military chiefs nowadays 'usually serve only as a channel through which the men in the laboratories transmit their views', the technician starts the crucial process of formulating the military requirement:

> It is he, the technician, not the commander in the field, who is at the heart of the arms race, who starts the process of formulating a so-called military nuclear need. It is he who has succeeded over the years in equating, and so confusing, nuclear destructive power with military strength, as though the former were the single and sufficient condition of military strength. The men in the nuclear weapons laboratories of both sides have succeeded in creating a world with an irrational foundation, on which a new set of political realities has in turn to be built. They have become the alchemists of our times, working in secret ways which cannot be divulged, casting spells which embrace us all.[2]

Zuckerman's thesis was reiterated and refined during the ensuing years. At the heart of it lies the proposition, not just that it is technicians, rather than politicians or the military, who are shaping a new and dangerous future, but that no-one is actually in command of events. 'It is all but impossible to believe that the process of defence R & D is under rational control.'[3]

Zuckerman cited a number of decisions on nuclear matters to

illustrate his claim. The first was the so-called Chevaline project to provide a British-built enhancement to Polaris warheads. 'It is now public knowledge', he wrote,

> that the men in the British weapons laboratory had set to work on the project, ... as far back as the late 1960s, without at the start seeking ministerial or any other approval.[4]

Another piece of evidence was that, before any decision had been made by the British government to replace Polaris with Trident,

> the men in the British nuclear weapons laboratory had pre-empted the situation. They had not only started to design a warhead for a MIRVed Trident missile; they had also, with American help, conducted underground tests of their designs.[5]

A final American illustration cited to show how weapons laboratories allegedly pre-empt strategic decisions is the cruise missile. It is not clear exactly what Zuckerman's contention is in relation to this particular development, but the main drift of his case is that cruise was originally developed, not in response to a defined strategic threat or need, but simply as a way of improving the survivability and postponing the obsolescence of the B-52 bomber.[6]

Claims of this kind had been made before. Publicly expressed concern about the role of scientists in weapons development goes back at least to President Eisenhower's farewell radio and television address to the American people in 1961. Eisenhower warned his audience 'that public policy could become the captive of a scientific-technological elite'.[7] In the intervening years, Herbert York, former Director of the Lawrence Livermore weapons laboratory and first Director of Defense Research and Engineering (1958–61), repeatedly explored much the same theme, notably in his book *Race to Oblivion: a Participant's View of the Arms Race*.[8] A number of academic studies, predominantly American, have attempted to explore the influence of scientists on weapons development, and looked at a selected number of case studies in this light.[9] Earl Mountbatten, experienced like Zuckerman in the ways of British defence decision-making, also argued that weapons innovation was driven more by the technical zeal of the scientists than by the needs of national defence.[10]

More recently the British Foreign Secretary Sir Geoffrey Howe, touched on the same theme in the speech to the Royal United Services Institute in March 1985. Warning of the danger of delaying consideration of the strategic implications of the Strategic Defense Initiative research programme, he said:

> Can we afford even now simply to wait for the scientists and military experts to deliver their results at some later stage? ... I do not believe so. ... research into new weapons and study of their strategic implications must go hand in hand. Otherwise, research may acquire an unstoppable momentum of its own, even

though the case for stopping may strengthen with the passage of years. Prevention may be better than later attempts at a cure. We must take care that political decisions are not pre-empted by the march of technology.[11]

Although many other 'insiders' have expressed views similar to Zuckerman's, it is his proposition which has attracted most attention. It is extensively quoted in Britain and North America, despite the fact that it is neither the most detailed nor the best documented formulation of the idea. It has achieved the status of 'the Zuckerman hypothesis'. The level of attention paid to its author derives partly from the prestige accorded to a former British Chief Scientific Adviser, privy to the most closely guarded secrets of one of the most secretive defence establishments in the world, sharing with the public some elements of his own privileged knowledge and purporting to expose the sensational irrationality within. His colourful formulation of the hypothesis, with its emotive reference to 'the alchemists of our times, working in secret ways that cannot be divulged, casting spells which embrace us all', touched on primitive fears of 'mad scientists' and of the unknown. It drew on a wider myth of the omnipotence of certain branches of modern science.

This myth stems from the perception that the implications of modern science in general and of nuclear physics in particular have transcended ordinary political processes. Key elements in the myth, as far as defence is concerned, are: the Einstein–Szilard letter which persuaded Roosevelt to embark upon the programme which eventually became the Manhattan Project; public reaction to the Hiroshima and Nagasaki bombings; the H-bomb project and popular accounts of the role played by Edward Teller in its conception; the rapid development of rocketry and guidance technology culminating in MIRVing and the American Apollo programme; and contemporary stories about the genesis and nature of 'star wars' research.

The issue which Zuckerman addresses is a vital one. It is essential to understand the processes by which new, high technology weapons are developed in order to devise effective defence, arms control or disarmament strategies. Zuckerman proposes a relatively simple model which, if it were essentially correct, would have major prescriptive implications for policy makers and popular movements alike. This chapter aims to explore what the Zuckerman hypothesis might actually mean, to examine its validity and usefulness, and to discuss some reasons for its wide appeal in Britain and the United States.

Zuckerman alleges that weapons scientists 'have created a world with an irrational foundation'. There are in fact two propositions here: weapons development is not under political control; and political control is rational control. This chapter concentrates on the first proposition, in the belief that weapons development ought to be under political control, whether or not that control can accurately be termed 'rational'. It draws on a number of case studies to illuminate the processes that actually do determine weapons development and the significance of the weapons scientist within them.

Zuckerman argues that it is the weapons scientists and R & D establishments which generate pressures for new weapons, that they initiate and lobby for research and development which is unamenable to proper political scrutiny and control, and that the resulting developments then fuel the arms race. This hypothesis is one of two distinct theories about the role of technology and the arms race; the other theory (which Zuckerman does not argue though he is commonly interpreted in this light) is that known as 'the technological imperative', a form of technological determinism. The technological imperative theory sees technology itself as an autonomous social force, generating an irresistable momentum. 'If it can be done, it will be.' An example of this theory is Ralph Lapp's book, *Arms Beyond Doubt: the Tyranny of Weapons Development*.[12] There are also elements of this theory, though somewhat confused with the rather different bureaucratic model, in Dietrich Schroeer's *Science, Technology and the Nuclear Arms Race*, which provides a crisp summary of it:

> Technology is one prime mover of the nuclear arms race through technological imperatives. Such technologies are so technically sweet and beautiful that they are difficult to resist, and the resulting technological progress may force some public policies that may not be desirable.[13]

The Zuckerman hypothesis is, like the theory of technological momentum, a supply-driven rather than a demand-led view of the arms race. However it focuses, not on the inexorable pressures of technology itself, but on the pressures generated by the weapons scientists, which broadly places it within what are termed organisational process or bureaucratic politics models of the arms race. Such models see weapons procurement policy as the outcome, not of deliberate government choice, but of the standard patterns of behaviour of the semi-autonomous and relatively permanent organisations involved in weapons development.[14] The four main groups identified by organisational process theorists have been: the military; the defence industries; government bureaucrats and politicians; and the weapons scientists. These groups are seen as internally competitive rather than homogeneous — the armed services competing with each other for shares of the budget, the defence companies competing for contracts, especially 'follow-on' orders, the politicians competing for votes, and so on. Whereas bureaucratic politics models usually locate the engine of the arms race in the systemic interactions, the shifting patterns of conflict and co-operation, among and between these groups, Zuckerman offers a more mono-causal analysis, focusing exclusively on weapons scientists.

The first issue is to work out exactly what Zuckerman means by saying that the arms race is fuelled by technicians in governmental and industrial laboratories. The term 'arms race' is itself an ambiguous one: the competition which it implies may be quantitative, qualitative or doctrinal, and may be of very differing degrees of intensity.[15] A number of analysts have concluded that it is not an appropriate description of the present 'arms evolution'.[16] This chapter does not go into that debate: it assumes that what Zuckerman is talking about is some kind of qualitative arms race and takes that to be an appropriate term, for the sake of the analysis.

In one sense, there is necessarily a connection between military R & D and any kind of qualitative arms race. The technological advances achieved by weapons scientists are a necessary condition for a qualitative arms race. Without the weapons scientists, such a race would certainly come to a halt. The relationship between military R & D and ever more sophisticated weapons development was described in this way by Mary Acland-Hood:

> While civil R and D can be regarded as a means of solving problems (and only creates them accidentally) military R and D is—in a sense—in the business of creating a virtually infinite series of problems. If techniques are found for detecting submarines, this will present the problem of designing submarines which cannot be detected. If ways are found of attacking missiles in their boost phase (while each is still one target), this will present the problem of finding ways of shortening the boost phase, and so on.[17]

Note that Acland-Hood's analysis implies that military R & D leads to two arms races, one external between rival powers and one internal between the designers of offensive and the designers of defensive systems. In the 'external' race, the principle of overreaction or 'worst case analysis', fostered by the secrecy of weapons development on all sides, generally guarantees that the reaction to a revelation of technological innovation in weapons usually exceeds that which was justified by the actual situation.[18] In this situation, defensive innovations are just as likely to fuel the arms race as offensive ones. Herbert York argued, in opposition to the anti-ballistic missile programme of the 1960s, that the programme was 'simply another step in the arms race. It represents a technical challenge to the technologists who design the offense. In designing around the ABM, these latter will usually come up with a more complex, more expensive, more deadly and more volatile offense.'[19] Twenty years on, the same case is often made against the Strategic Defense Initiative: a bigger and better illustration of the 'fallacy of the last move'.[20]

Zuckerman—unlike Acland-Hood—implies that military R & D is a sufficient condition for the arms race. That is, that the political control which is supposed to be exerted, to promote stabilising technologies and to prevent destabilising ones, is not in fact operating. Zuckerman's hypothesis, true to its bureaucratic model, speaks not so much of military R & D in the abstract as of the scientists and technicians in the weapons laboratories. The second issue which arises is to explore who exactly these may be, and how many of them there are; that is, how large and potentially influential is this interest group?

We are certainly considering very substantial numbers. It is estimated that about one quarter of the world's scientists and engineers involved in research are employed in military R & D[21] and are heavily concentrated in the United States, Europe and the Soviet Union. In both the United States and the United Kingdom the greater part of military R & D is spent in industry, but the numbers of scientists employed there are difficult to establish. Figures are, however, available for those in direct government employment. In the United States, where one quarter of the total military R & D budget is spent directly by the Government, 14,279 scientists and 55,532 engineers were directly employed by the Department of Defense in September 1977.[22] In the United Kingdom, some 4,800 qualified scientists and engineers were directly employed by the Ministry of Defence on military R & D in 1984–85.[23]

Then there is another significant category of weapons scientists — not specifically mentioned by Zuckerman — academics working in university laboratories or, in some cases, in federal contract research centres. In the United States, some 3 per cent of the total defence budget is spent by universities and 2 per cent by federal contract research centres (such as the Lawrence Livermore Laboratory) (1983 figures). Even allowing for their much smaller number, university scientists in the United Kingdom are correspondingly less involved in military R & D. There is no British equivalent to the government-sponsored university laboratories of the United States, and very few university scientists are involved as consultants on military projects. It is estimated that only £13 million out of a total of £159 million of net extramural military research was spent in universities and further educational establishments in 1986–87.[24]

But none of these figures necessarily does justice to the importance of university scientists in military research. Academic scientists tend to make up a disproportionate number of the more talented of all weapons scientists (who themselves contain disproportionately the more talented of all scientists). Defence-funded research offers tempting professional opportunities and high material rewards, especially in the United States. President Eisenhower, in his farewell address, drew special attention to the teams of university scientists working on defence contracts and to the effect of defence contracts on their research work:

> the free university, historically the fountain head of free ideas and scientific discovery, has experienced a revolution in the conduct of research. Partly because of the huge costs involved, a government contract becomes virtually a substitute for intellectual curiosity. . . . The prospect of domination of the nation's scholars by federal employment, project allocations, and the power of money is ever present—and is gravely to be regarded.[25]

So university scientists working on defence contracts, as well as those directly employed in government defence laboratories and by defence contractors, have to be taken into account.

Another category of scientist, not specifically mentioned by Zuckerman though highly relevant to his hypothesis, is that of scientific bureaucrats. They do not themselves directly spend military R & D but have much power over its distribution. Scientific bureaucrats are numerous in the United States. The scientists on the Presidential Scientific Advisory Council (PSAC) or those in the Defense Advanced Research Projects Agency (DARPA) are perhaps the most influential and best known of what is almost a Washington industry. There is barely a similar category in existence in Britain where the Chief Scientific Adviser (as chapter 6 makes clear) operates largely in isolation from the broader scientific community. In the United States there are substantial numbers of other scientists, not included in any of the above categories, who are in fact engaged on defence-related research, although it is not financed directly from the defence budget. For example, the budgets of the Department of Energy and the National Aeronautics and Space Administration finance a substantial contingent of military-related researchers. Some military research and development is also known to have been hidden under other appropriations, such as that of the Central Intelligence Agency.[26]

This breakdown of the different categories of scientists' employment should not, of course, be allowed to obscure the fact that individual scientists may move frequently from one type of employer to another. During the early years after the founding of the Advanced Research Projects Agency (ARPA) within the Department of Defense in 1958, for instance, there was an especially rapid interchange of key technical personnel between ARPA and industry.[27]

The Zuckerman hypothesis lumps all weapons scientists together, as a single, relatively coherent group with particular responsibility for generating an arms race. It could, in fact, be the case that the very different types of organisations for which weapons scientists work create significant differences between them. Indeed, organisational process models usually see each actor's stand on an issue as greatly influenced by his or her organisational affiliation: 'where you stand depends on where you sit'.[28] In that case, if weapons scientists do indeed generate pressures for weapons development, that pressure could more appropriately be considered as a subset of wider industrial or bureaucratic pressures, with industrial scientists being part of industrial lobbying, military scientists part of Department of Defense and service pressures, and so on. Zuckerman's hypothesis implies that what all these scientists have in common is more important than what divides them, and what they have in common is distinctively responsible for fuelling the arms race.

How, according to Zuckerman, do these scientists control weapons development? His hypothesis stresses their role as the generators of ideas. There can be no doubt that weapons scientists generate ideas; that is part of their function. But what Zuckerman suggests is that, rather than using their scientific creativity to provide answers to predetermined strategic problems—as is supposed to be the case—weapons scientists, in fact, generate and successively promote new technological ideas which subsequently obtain *post hoc* strategic rationalisation. In other words, they produce solutions in search of problems. This means, in effect, according to Zuckerman, that weapons scientists, albeit without military training or experience, begin the process of formulating military requirements and use military chiefs merely 'as a channel for the transmission of their views to Government'. It is the way weapons scientists usurp the strategic function of beginning the weapons development process, without adequate political scrutiny or control in the early stages and unamenable to it later, that is the burden of Zuckerman's case.

It is necessary to look elsewhere than to Zuckerman for more detailed analysis of how scientists unduly influence weapons development. Some analysts have, for instance, emphasised the characteristic ethos of professional scientists, as innovation-oriented. Whereas military institutions, the armed services and defence bureaucracies—conscious that the technologies embodied in new weapons systems may entail disturbing new forms of social organisation[29]—tend to be conservative, the pushing out of the frontiers of the new and unknown is the research scientists' lifeblood. Warner R. Schilling, reporting on the comments of several (unnamed) scientists he had spoken to who had been closely involved in the American development of the H-bomb, said that:

> It is their feeling that there are times when the technician does take over, that when the scientist is faced with an interesting and challenging problem his inclination is to get to work on it, and that under these circumstances he should not be the first person to be expected to keep larger policy considerations in balance.
>
> This predisposition, 'technology for its own sweet sake', appears to have its roots in two more of science's central credos: the belief in the value of pursuing knowledge for its own sake and the belief that the best motivation for the direction of research is the strength and character of individual curiosities. But the direction and strength of scientific interests and curiosities is not necessarily coincident with the requirements of military and foreign policy.[30]

To interpret progress in scientific terms alone can be contagious. Herbert York argues that many weapons scientists and engineers came by the end of the fifties to believe that the normal and desirable technological state of affairs was one of a continuing flow of new scientific discoveries. 'They virtually promised their military and congressional supporters that the future would be as glorious as the recent past, only more so.'[31]

Others have stressed the central importance of scientific rivalry. The influential American strategist Albert Wohlstetter believed, like Zuckerman, that scientists influenced the development of strategy, in a way which they were not qualified to do. He emphasised the commitment which scientists feel to their own ideas:

> Politicians and military men have all at one time or another allowed their fondness for some pet scheme or device to lead them into wishful negligence of some obvious counter. And while some scientists have been both realistic and ingenious in this respect, there are a great many cases where they too have been wishful. When a man concentrates for years on a technically brilliant design of a warning system against a specified, rather straightforward sort of attack, he might very well be aggrieved at the specter of any enemy who simply flied [sic] around or over or under it, or worse, flies into it so frequently in time of peace as to destroy much of its utility in warning of the outbreak of war.
>
> On my own observation, I would say that it is a rare scientist (or for that matter a rare anybody) who can with equanimity wear both of two hats: design a system well and then wholeheartedly knock it down; even as a preface to building it up again in a way that will sustain the shocks. More frequently you can find a scientist extremely ingenious in thinking up counters to a policy on his own side *rivalling* the one he is recommending. Then he can work up a little enthusiasm. (Then, in fact, he may be carried away, and devise for his rival a hypothetical enemy who is not merely ingenious, but hardly subject to any realistic limitation.)[32]

It has been suggested, as for instance by Acland-Hood earlier, that this kind of internal scientific competition leads to actions and reactions, not so much *between* the power blocs, as *within* them. The early developments of MIRV and ABM, for instance, were, in York's view, 'largely the result of a continuously reciprocating process consisting of a technological challenge put out by the designers of our defense and accepted by the designers of our offense, then followed by a similar challenge/response in the reverse direction.'[33]

Others have attached importance to the counter-intuitive effects of budgetary control systems in giving undue influence over weapons development to research scientists. The American academic Harvey Brooks, for instance, is, like Zuckerman, convinced that weapons research and development often generate irresistible pressures to get the resulting hardware into production, without adequate political control being exerted. His extended illustration goes beyond Zuckerman in hinting at the differing roles of the military–technical community and the industrial–technical community. Although his example is American, the phenomena of *post hoc* rationalisation and hidden agendas in bidding for resources are hardly unknown elsewhere:

> Congress is often reluctant to appropriate research funds without specific plans for a weapons system, thus precluding research to generate options from which to select. Too often development is treated as a part of a rational process in which the final result should be fully defined in advance. Research in industry is also funded with the strong expectation that the results will be converted into an operational weapons system. Furthermore, in a period when force levels are relatively stable, service technical organisations and their industrial clients can guarantee their own survival only by generating a continual stream of qualitative improvements to make existing deployments obsolete. This tends to be the case even though, in fact, the majority of systems undergo some development, but never reach operational deployment. The system also tends to favor 'product differentiation' for its own sake, much as in other sectors of the industrial economy. Research and development that do not lead to a deployed system are looked upon as failures and a waste of resources; little value is attached in practice to the knowledge gained in a project, if there are no tangible results.
>
> The process is driven by a 'military requirements' system that is partly based on fiction. Experience has taught the military technical community that it is much easier to sell interesting research if it can be pushed as a fully conceptualised weapon system meeting a well-defined threat from a postulated opponent. In practice, both the threat and the requirement may have been invented to provide a rationale for a development program started for other reasons, such as to perpetuate existing organisations, or to exploit a 'sweet' technical concept.[34]

So much for other analyses of how scientists may exert undue influence over weapons development. How can we now begin to assess the usefulness of the Zuckerman hypothesis?

Criteria for the assessment of usefulness

The Zuckerman hypothesis that scientific pressure groups are the motive forces behind the arms race has some superficial plausibility. The early development of atomic weapons seems to support it. It was scientists, mainly refugees from central Europe, who first saw the military possibilities of creating a self-sustaining, nuclear fission chain reaction and who lobbied the wartime American administration for funds for a military research programme.[35] Similarly, after the war, it was scientists who were anxious to explore what they believed to be the dramatic potential for nuclear development beyond the crude atomic bombs dropped on Hiroshima and Nagasaki. In contrast the military, professionally conservative in this as in other instances, thought only in terms of incremental improvements to the weapons they already had. As Edward Teller described in *The Legacy of Hiroshima*,[36] scientists at Los Alamos in the late 1940s found their situation frustrating. Some of them wanted to get on with new and different designs (as a minimum, designs which were different in terms of weight, explosive yield and general dimensions), but the armed services were having difficulty imagining what they would do with anything different from the 'Little Boy' or the 'Fat Man', dropped on Hiroshima and Nagasaki and designed specifically to fit the B-29 bomber.[37]

Once the Soviet Union had developed fission, and subsequently hydrogen, weapons, it seems plausible that the United States' emphasis on staying ahead in a technological race with the Soviet Union, and on counterforce strategies which demand continuing qualitative improvements in weapons performance, would give power to scientists and technologists. We shall be looking in detail later at the role of scientists in the development of the hydrogen bomb.

However, the implication of Zuckerman's hypothesis is not just that there are one or two cases in which scientists were the motive force of undesirable new weapons developments, but that this is *characteristically* their distinctive role. What is crucial to it is the idea that proper constitutional control over weapons development by the political process is characteristically pre-empted; that weapons scientists are responsible for what Eisenhower called 'the disastrous rise of misplaced power in the process of acquisition of modern arms'.[38] The hypothesis is not a scientifically rigorous proposition in the sense that it could be said to be falsified by one contrary example. The Zuckerman hypothesis is clearly a generalisation rather than an invariable rule, and would have to be shown to be *characteristically* inadequate to invalidate it as a useful general proposition.

Quite how to test the Zuckerman hypothesis against historical experience is itself a complicated question. It is unclear what evidence could invalidate it. It might be possible to demonstrate that, for each significant advance in weapons technology, some weapons scientists who were involved in its development argued against its deployment on the grounds that the new technology would be unnecessary or destabilising. Robert Oppenheimer's opposition to the development of the H-bomb would seem to be an instance of this, as would Peter Hagelstein's in relation to third generation nuclear warheads and the Strategic Defense Initiative.[39] However, even if such instances were characteristic, the Zuckerman hypothesis would not necessarily be falsified.

This is partly because the research and development of a new technology or weapons system alone, without deployment, may fuel an arms race. It may be sufficient to rouse a 'worst case' reaction in a potential adversary or even to stimulate another branch of the state's military R & D establishment to develop countermeasures. Harvey Brooks, for example, has drawn attention to the former possibility:

> The testing or demonstration of new technological weapons can be politically as destabilising as their actual deployment, especially in the absence of reliable

intelligence. Thus the projected bomber gap of the mid nineteen-fifties and the alleged missile gap of the early sixties were inferred from Soviet capabilities that had only been demonstrated on the level of research and development or prototype testing. The inferences proved unfounded, since there was no major deployment of either bombers or ICBMs. But the United States had in the meanwhile launched a major build-up of bombers and ICBMs, which gradually acquired a justification of its own, independent of the hypothesised but non-existent threat that had originally inspired it.[40]

So such a strategy for testing Zuckerman's claim that the activities of weapons scientists provide 'at base, the momentum of the arms race' would need to show that weapons scientists have also attempted to stifle potentially destabilising *research*, not just testing and deployment. Halting research on new weapons is a particularly difficult issue, because ideological reasons against it can readily be found. Mrs Thatcher defended the Strategic Defense Initiative programme to Mr Gorbachev in March 1987 on the grounds that 'You cannot stop such research any more than you can stop the onward march of science in general'.[41] In any event, for scientists themselves to try to stop scientific research runs counter, not just to possible immediate self-interest, but to the whole weight of the scientific ethos. Brooks, writing in 1976, concludes that:

> Until recently, official scientific advisers and some of the scientific community interested in such matters have advocated research and development to improve certain weapons technologies, even when they have strongly opposed the deployment of the corresponding specific weapons, whether because they were ineffective or because they would accelerate the arms race. Over a period of nearly fifteen years, independent scientific advice discouraged deployment of successive ABM systems. But each time the independent scientists recommended against deployment, they also advocated more research aimed at the next level of improvement, primarily on the grounds that we could not afford to be taken by surprise over what was technically possible.[42]

However, even if it could be demonstrated that some weapons scientists had opposed the research, development, testing and deployment of every significant innovation in military technology, the Zuckerman hypothesis could still stand. For it is the position taken by the scientific institutions and the activities of the majority of the scientists involved that is most relevant, not the attitudes of untypical individuals, however prominent they may be.

In this chapter we do not seek to challenge the view that the weapons laboratories tend to pursue their institutional interests nor that scientists involved in military work are inclined to follow intriguing or promising technological developments irrespective of overall strategic requirements. This view seems plausible and would in any case be difficult to investigate comprehensively. The processes of weapons development, particularly in their early stages, are both complex and

The Alchemists of Our Time

among the most secret of a state's activities. The very complexity and secrecy which leads to doubts about political or institutional control being exerted over R & D complicates the task of academic enquiry.

However, we do aim to investigate the *primacy* of the role given to scientists and scientific institutions in explanations of the development of new weapons systems. We examine a number of cases where the Zuckerman hypothesis seems most plausible to see whether the hypothesis emerges as a useful and accurate generalisation. The test of the hypothesis is therefore not whether or not scientists and scientific institutions have acted as an interest group in the development of the new technologies, but rather whether focusing on the role of this particular interest group provides a seriously misleading image of the way in which the weapons systems have evolved.

As observed earlier, histories of the Manhattan project, and of the key role played by scientists in the project, have tended to enhance the popular appeal of the Zuckerman hypothesis as a central explanation for the arms race. Yet a moment's reflection shows that the hypothesis has little relevance in this case. Szilard, Einstein and their scientific colleagues drew the attention of political authorities to the possibility of developing an atomic bomb, not because they were fascinated by the intriguing scientific problems nor as a result of sectional interests, but because of a clear awareness of the danger that Nazi Germany might develop such a weapon. Vast resources were allocated to the project as a result of definite political decisions and the development of the bomb was closely supervised by military and political authorities.

Of course, the vast military R & D establishments have emerged only since the end of World War II. So the Zuckerman hypothesis might be expected to have more validity for weapons development since 1945. We therefore examine the development of the hydrogen bomb, MIRV technology, the MX missile, the strategic cruise missile and the British Chevaline project—all cases where the Zuckerman hypothesis appears to be most plausible—in order to assess its explanatory value.

Truman's H-bomb decision

On 31 January 1950, President Harry S. Truman made the following statement:

> It is part of my responsibility as Commander-in-Chief of the Armed Forces to see to it that our country is able to defend itself against any possible aggressor. Accordingly, I have directed the Atomic Energy Commission *to continue* its work on all forms of atomic weapons, *including the so-called hydrogen or superbomb*. Like all other work in the field of atomic weapons, *it is being* and will be carried forward on a basis consistent with the overall objectives of our program for peace and security.[43]
> [Emphasis added.]

This statement represented the outcome of several months' intensive policy arguments within the American administration and the scientific community. It also marked the formal beginning of the American effort to develop a hydrogen bomb or fusion weapon. But the President's statement, as delivered, differed significantly from earlier drafts, in that the italicised passages refer to work under way on fusion weapons as if it were nearing completion. An earlier version of the statement drafted by a special committee of the National Security Council formed to advise the president on whether an H-bomb programme was desirable had spoken only of continuing 'with the development of all forms of atomic weapons. This work includes a project looking toward a test of the feasibility of the hydrogen bomb'.[44]

It is not clear why the draft was strengthened in the way it was. It may be that the President wished to give the impression that the United States was closer to the fusion bomb than was actually the case. Whatever the rationale for Truman's formulation, however, there is no dispute that it misrepresented the state of American theoretical and practical knowledge about a hydrogen weapon. Even the bomb's most enthusiastic advocate, Edward Teller, recorded his amazement at the tone of the President's announcement. The impression was being given, he wrote,

> that we could produce a hydrogen weapon by tightening a last few screws. People understood from the pronouncement that the job was almost done. Actually work had not begun. We had eight years of thermonuclear fantasies, theories and calculations behind us; but we had established no connection between theory and reality.[45]

Teller's amazement was widespread in the scientific community. For example, Norris E. Bradbury, the first postwar director of the Los Alamos Laboratory said:

> The state of knowledge of thermonuclear systems during the war, and thereafter, and really up until the spring of 1951, was such as to make the practical utility or even the workability in any useful sense of what was then imagined as a thermonuclear weapon extremely questionable.[46]

Despite this, the Truman statement initiated a massive, concerted research and development effort which culminated in the first test of a rudimentary fusion weapon on 1 November 1952.

The H-bomb story is a good test of the Zuckerman hypothesis. In the first place, the H-bomb represented a significant technological innovation. Although it obviously depended on existing technology, it was not merely an extrapolation from it. Secondly, the innovation itself had a significant impact on the nuclear arms race. And thirdly, the chronology and main phases of the decision-making process are relatively well known.

Fusion technology

The theoretical history of fusion weapons stretches back to the work of Hans Bethe on thermonuclear reactions in stellar interiors in the early 1930s. In the early years of the Second World War there was some research on thermonuclear reactions in the United States by a number of physicists—notably Edward Teller, Enrico Fermi and Emil Konopinsky. When a summer-long conference was convened in Berkeley in 1942, ostensibly to discuss plans for the explosive design for an atomic (i.e. fission) weapon, much of the discussion was about the theoretical possibilities of a hydrogen bomb.[47] However, when the Los Alamos Laboratory was set up later that year, the summertime theorising about fusion reaction was relegated to low-priority status to make way for the rapid development of a fission bomb. This was partly because the development of the latter turned out to be more difficult than had originally been assumed, and partly because a fission weapon was, in any case, a practicable prerequisite for a fusion one.

This is because the start of a fusion reaction requires very high temperatures, and the only practicable way of achieving these on earth was, until recently, by means of a fission explosion. The implication is that a fusion weapon must be at least a three-stage device. In the first stage a conventional explosion is detonated. This then initiates a fission reaction, i.e. an atomic bomb. In the third stage the energy from the fission explosion initiates a fusion reaction in the hydrogen nuclei. In most thermonuclear warheads there is also a fourth stage, in which neutrons from the fusion reaction initiate additional fission, on a scale vastly greater than the first, in a surrounding blanket of fissionable material such as uranium-238.

There are, in principle, few limits to the size and yield of a thermonuclear weapon, and some weapon designs have provided about 1,000 times more energy than the fission process on which the early atomic bombs were based. Thermonuclear weapons thus represented a quantum leap not just in technology, but also in their significance as weapons.

With the attention of the Los Alamos Laboratory focused almost exclusively on the fission bomb, virtually the only researcher actively studying the hydrogen bomb in the period up to 1946 was Edward Teller. He ran into some problems and later recounted that

> by the end of the war the question whether a fusion bomb could be made or not was completely up in the air.[48]

After the Hiroshima and the Nagasaki explosion, the Los Alamos Director, Robert Oppenheimer, made it clear that the small amount of work being done on the hydrogen weapon (or 'superbomb', as Teller called it) should be discontinued. Oppenheimer informed the government that the scientists under his control would prefer not to work on the weapon unless ordered to do so on grounds of national policy. Between the giving of this advice in 1945 and the initiation of a concerted H-bomb programme in 1950, as a result of Truman's decision, very little work was done on thermonuclear weapons. Few physicists devoted more than infrequent attention to the application of fusion theory to a practicable weapon. The most notable exceptions — Teller, Nordheim and Richtmeyer — studied only specific fusion problems, with 'very little progress toward either a feasible theory or a workable engineering approach'.[49] Accordingly the President's decision to embark upon the development of the Superbomb represented a leap into the unknown.

This explains why there was surprise in the scientific community at the implicit suggestion in the Truman statement of January 1950 that research on the development of a hydrogen bomb was well advanced. In fact, the reverse was the case. As the Columbia physicist Isidor Rabi, a member of the General Advisory Committee of the AEC wrote:

> It was a very difficult question, because here is a statement from the President to do something that nobody knew how to do.[50]

It is clear therefore that Truman's decision to develop the H-bomb was not a result of simple 'technological push' or the rubber-stamping of work already embarked upon by scientists. The Zuckerman hypothesis would suggest that the decision may have been a result of pressures from weapons laboratories; yet, as we have seen, the latter were showing little interest in developing the H-bomb — indeed, they were arguing against it. Edward Teller, sometimes referred to as the 'Father of the hydrogen bomb', was undoubtedly a powerful advocate for the weapon. But his views were untypical of those of the scientific community. If he and a few colleagues found a ready audience among important military and political decision-makers, this cannot be explained by reference to the influence of scientific institutions, but rather as a consequence of the interaction of international events, political pressures or broader bureaucratic interests. And indeed, history shows this to be the case.[51]

The political and bureaucratic contexts in which the H-bomb decision was made were governed by the post-war development of a 'cold war' between the United States and the Soviet Union, and by speculation of how long it would take the Soviets to acquire nuclear weapons. Within the American administration there was conflict between the line advocated by George Kennan and the State Department (which essentially argued that Soviet expansionism should be countered with political and economic measures combined with only limited military forces) and that of the Joint Chiefs of Staff (JCS) and others who feared that this line seriously underestimated the Soviet military threat and would undermine service requests for military hardware and manpower. The JCS plan called for an enhanced air-power programme, with the aim of providing the United States with the power to carry out a massive strategic air strike in a crisis. In the worsening international climate, the JCS approach became dominant, and led to a comprehensive military rebuilding programme in the United States. In the Soviet Union the same developments led to an intensification of research on its atomic bomb.

The decision

In 1947 no usable system for monitoring Soviet atomic tests by detecting radioactive debris in the atmosphere was available. Eisenhower ordered the development of an effective monitoring system as soon as possible and a rudimentary system was in place by the summer of 1949. Towards the end of August, American B-29s picked up increased levels of radiation over the Pacific, the source of which was eventually identified as somewhere in Soviet Central Asia. By mid-September, the United States intelligence organisations were certain that the Soviet Union had exploded an atomic weapon between 26 and 29 August. Truman announced the fact to the American public on 23 September.[52]

In the United States, the Soviet explosion stimulated a major inter-agency debate that lasted for several months. The main actors were the scientific advisory elite clustered round the Atomic Energy Commission (AEC) and the Los Alamos Laboratory; the national security staffs at the Departments of State and Defense; a number of politicians, notably Senator McMahon, Chairman of the Joint Committee on Atomic Energy (JCAE); and the civilian commissioners on the AEC. It is not known what role, if any, the intelligence agencies played.

As described by Stein, the debate began at the AEC. Commissioner Lewis Strauss, prompted by a member of the National Security Council, suggested on 5 October 1949 that the AEC should report to the President on the feasibility and desirability of a hydrogen bomb programme. The AEC then formally asked its scientific advisory group, the General Advisory Committee (GAC) whether a crash programme to develop an H-bomb was advisable in the light of the Soviet atomic test.

GAC members were initially divided about this. Oppenheimer (then the GAC's chairman) was against, but Teller, E. O. Lawrence and Luis Alvarez were more pessimistic about Soviet intentions and capabilities.

On or around 6 October, Teller invited the other two, together with Ulam and a Soviet émigré, George Gamow, to his home to discuss Soviet interest in the possible applications of fusion research. Gamow told an anecdote about an approach he had had from a very senior Soviet minister which persuaded the assembled company that the Soviet administration believed that serious fusion research was already under way in the United States (with the implicit corollary that it would therefore be prudent for the USSR to follow suit). This was, of course, pure speculation. No one present had any means of knowing whether their suspicions or fears were well founded.

On 10 October, Lawrence, Alvarez and a colleague met two politicians who were to play a significant part in the story. One was Carl Hinslow, a Representative from California. The other was Brien McMahon, a Senator from California and chairman of the JCAE. From the outset, it appears that McMahon was a convinced advocate of the H-bomb.

On 14 October, the military enter the story. The JCAE met the JCS. At the meeting the most senior air officer advocated a programme to develop the H-bomb. This meeting may have been briefed by some scientists. At any rate, there is a letter dated 21 October from Robert Oppenheimer to James Conant of Harvard University reporting that Lawrence and Teller had made a significant impact on the JCAE and the JCS.

Formally, however, at this stage the responsibility for advising the President rested with the AEC and its GAC. Three of the AEC's five commissioners (Lilienthal, the chairman, together with Sumner Pike and Henry D. Smyth) were opposed to any immediate attempt to develop a hydrogen weapon. The remaining two commissioners (Lewis Strauss and Gordon Dean) were already moving to a position of favouring an H-bomb programme.

The GAC held its first meeting to consider the issue between 29 October and 1 November. Eight of the nine members of the GAC attended. The Committee's brief was to give a scientific opinion on the *feasibility* of a hydrogen weapon. Despite these limited terms of reference, however, the GAC consulted a wide range of officials in the four days, among them George Kennan of the Policy Planning Staff, the chairman of the JCS, the Secretary of Defense and some intelligence personnel. According to Stein,[54] the GAC was aware of Teller's lobbying activities, and most if not all of its members were opposed to the hydrogen weapon. In the end, the GAC recommended against high-priority development of the 'Super' on a variety of scientific and non-scientific grounds. The Committee claimed that since no one had yet a clear idea of how to go about building such a weapon, it would be premature to discuss a crash programme. Reactor capacity would have to be increased before the required temperatures and pressures could be combined with the proper raw materials. The report giving this opinion was signed by six scientists. Two more (Fermi and Isidor Rabi) wrote a minority appendix opposing the Super in even stronger terms. From this point until the President's decision, the GAC advice seems to have adhered to this general line.

Some time in early November, the AEC commissioners met the JCAE. The Chairman of the JCAE, Senator McMahon, later revealed to Teller that the GAC's recommendations had 'made him sick'. On 3 November the Policy Planning Staff met to discuss the Super. At this stage, Acheson, the Secretary of State, still seems to have been sceptical. He suggested that an 18- to 24-month moratorium on thermonuclear research development—'bilateral if possible, unilateral if necessary'.[53] Kennan was resolutely opposed to the bomb. The following day (4 November), the AEC discussed the Super, though apparently inconclusively. On 7 November Lilienthal went to the White House to offer his resignation as planned. Truman asked him to stay in post until the thermonuclear issue was sorted out and Lilienthal agreed. Two days later, he wrote to the President arguing against the Super. At this point, the AEC still seems to have been divided three to two against the bomb.

On 15 November, Senator McMahon went to Los Alamos to be briefed on the bomb, and was told by Teller that the Super had more than a 50 per cent chance of success.[54] Four days later, Truman authorised the establishment of a special committee of the NSC to advise on the H-bomb issue. The membership of this *ad hoc* committee was the Secretaries of State and Defense, plus the Chairman of the AEC. The first thing the committee did was to set up some working groups of officials. These groups began work on 28 November. Seven days earlier, McMahon wrote a personal letter to Truman arguing that to allow American capability to stagnate at the level of fission weapons would

> place a ceiling on our military advancement; for I do not know how the Los Alamos Laboratory would occupy itself, after a few years have passed—unless it ventured far into the thermonuclear field.[55]

On 25 November, Truman received a similar message from AEC Commissioner Strauss. On 3 December, the GAC met again and its members reaffirmed their opposition to any crash programme on the Super. This view was also passed to the special NSC committee.

On 22 December the NSC committee met and had an inconclusive discussion. Acheson was still formally uncommitted to the Super, though in deepening political trouble because of his support for Alger Hiss. In fact Stein believes that the Secretary of State had by this time decided privately to endorse the new weapon. A few days later,

Lilienthal met General Bradley (chairman of the JCS), but failed to persuade him from his advocacy of the Super.

After Christmas the pace of events quickened. George Kennan departed to the State Department as Special Counsellor, and was replaced at the Policy Planning Staff by Paul Nitze who advocated (to Acheson) proceeding with a feasibility programme on thermonuclear weapons. The JCAE was briefed on 18 January by an AEC technical specialist on delivery methods and tritium production, and by General Bradley on 20 January. During one of these meetings the JCAE chairman berated the GAC for exceeding its competence. All of these meetings took place against a background of an increasing trickle of inspired newspaper leaks about the Super.

A key date in the story seems to have been 27 January. On that day Lilienthal argued the GAC case to the JCAE, while Truman admitted in a press conference that the H-bomb was under active consideration and the President was informed of Klaus Fuchs's arrest in England, and of the likely implications of his alleged espionage. Four days later the NSC special committee met for the last time and agreed to a cautious recommendation to proceed; this was subsequently strengthened before being transmuted into Truman's fateful announcement. The H-bomb issue had been decided: the Super was under way, even though nobody with responsibility for these matters had any idea whether the project was feasible or not.

Discussion

What are the salient features of the H-bomb story from the point of view of the Zuckerman hypothesis? One is that it was not clear at the time the decision was made whether what was being proposed was scientifically feasible or not. Another is that the majority of the responsible scientific advisers were opposed to the programme from the outset, and did not waver in their opposition. Granted there was some dissension in the scientific camp, and Teller clearly made good use of his political contacts and his own powers of advocacy. But it seems inconceivable that Teller and his scientific allies would have made much progress unless powerful figures in the political and bureaucratic systems were already strongly inclined towards the Super.

It is significant that, although international events and political and bureaucratic concerns appeared to dominate the decision to develop the H-bomb, two key actors—Truman and McMahon—chose to exaggerate the role of scientists and scientific institutions in their public advocacy. Truman falsely gave the impression to the public that scientific work on the H-bomb had already progressed much further than it in fact had. McMahon argued that the H-bomb project was necessary for the maintenance of the design teams at the nuclear weapons laboratories — a case that was not made by the leaders of these scientific institutions themselves. Politicians were thus, for their own political purposes, giving a misleading emphasis to the significance of scientific institutions in the shaping of the H-bomb decision. This should act as a warning against accepting statements lending credence to the Zuckerman hypothesis at their face value.

The case of MIRV

Another much-cited example in this context is MIRV, a technological breakthrough not self-evidently pre-specified by strategic requirements and subsequently widely considered to be destabilising. Herbert York and Graham Allison among others have undertaken studies of MIRV in this context:[56] what follows is a further analysis of the development of MIRV, to try to examine the explanatory value of the Zuckerman hypothesis in this case.

MIRV (multiple independently targeted re-entry vehicles) is the acronym used to describe the dominant technology of current generations of ballistic missile systems. The decision to MIRV is widely regarded as one of the milestones in the history of the nuclear arms race. Ted Greenwood, for example, in his study of this technology described MIRV as 'probably the single most important technical innovation in the field of offensive strategic forces since the ballistic missile'.[57] Its perceived importance stemmed from several factors. One is that MIRVing greatly increased the numbers of independently targeted, strategic nuclear warheads available to both sides. A second factor is that, by increasing the probability that a nuclear strike would inflict significant damage, MIRV technology increased the incentives for pre-emptive strikes in a time of crisis. Finally, MIRVing greatly complicated the (already formidable) problems of anti-ballistic missile defence.

It has been alleged that the decision to develop MIRV is a classic illustration of the way in which technological innovation drives nuclear doctrine. York, for instance, has argued that:

> Almost all the important decisions were technologically determined.... Strategic analysis entered only fairly late (and indecisively) in a relatively narrow argument over the relationship of MIRV to the arms race. More general strategic thinking, and political considerations did not enter into the process until it was too late for them to have any effect.[58]

The purpose of this section is therefore to provide a summary account of the MIRV decision, following which we shall examine the implications of the story for the Zuckerman hypothesis.

For our purposes, the history of ballistic missile development can be divided into three phases. In the first phase each missile carried a single warhead, which constituted its re-entry vehicle. In the second phase the physical size of warheads was reduced to the point where it became feasible to pack several warheads into the payload of a single missile. This is known as multiple re-entry vehicle (MRV) technology. The first multiple warhead to be developed in the United States was the Polaris A-3, which carried three re-entry vehicles each with a yield of 200 kilotons. After the A-3s booster burned out, the three re-entry vehicles were separated by mechanical means into a triangular pattern centred on the target. But this separation was not adjustable, so that the 'footprint' on the target was a function of the flight distance. The system was designed so that at maximum range the separation would not be greater than the extent of a city. The A-3 was authorised in 1960, first tested in 1962 and became operational in 1964.

MIRV technology (the third phase) differs from MRV in one fundamental respect, namely that MIRVed vehicles are controllable in the post-boost phase. A manoeuvrable final stage of the missile, called a post boost control system or, more commonly, a 'bus', carries the re-entry vehicles system. The bus can change velocity, orientation and therefore its trajectory. After the main missile booster has burned out and dropped away, the rocket motor on the bus ignites, corrects the trajectory and then drops off the re-entry vehicles one by one, making adjustments and corrections as it goes. The number of re-entry vehicles in deployed systems ranges from three (Minutemen III) to 14 (Poseidon).

The basic advantage of a MIRV system is its ability to deliver accurately several warheads along different trajectories. This independent targeting capability can be used in several ways to enhance the destructiveness of a weapons system against a single target. Destructiveness can be optimised, for example, by careful distribution of MIRVed warheads with respect to a large target. A missile defence system can be overwhelmed by the sheer numbers of incoming MIRVed vehicles. MIRVs can also be used to attack different targets (subject to limitations on the amount of separation that can be achieved by the system). For Poseidon, for example, the maximum down-range separation at a range of 1,800–2,000 nautical miles is 300–400 miles (the cross-range separation, that is, the separation in a direction at right angles to the flight path, is about half that).

The account of MIRV development which follows is mainly derived from Greenwood's book, *The Making of MIRV: A Study in Defense Decision Making*.[59] In 1957, shortly after the launching of the first Sputnik, the American Department of Defense set up a committee (the Re-Entry Body Identification Group), to study the question of whether designers of missiles should take anti-ballistic missile systems seriously, and if so, what to do about them. There were two considerations in mind here: to determine what challenges the American designers of ABM systems might have to face from 'penetration aids', and to identify techniques to help American offensive missiles evade a possible future Soviet ABM system. It is important to note that ABMs were not then currently deployed by the Soviet Union—what the Department of Defense was doing was not *responding* to an actual Soviet threat but *anticipating* one. This has obvious relevance to the thesis that, research, or suspected research, may alone stimulate the arms race.

The committee described a number of possible counter-measures or penetration aids such as decoys, chaff, reduced radar cross-sections, and multiple warheads to overload such systems. The Air Force incorporated several deception devices, such as decoys, and the Navy developed a multiple warhead system for the A-3 version of Polaris. At the time the A-3 was authorised, in September 1960, the Ballistic Missile Division of the Air Force Systems Command was studying new re-entry vehicles for use on the land-based Minutemen, Titan II and Atlas/Titan I forces. Several mechanisms for releasing multiple warheads from these missiles were considered, but in the end the only one considered for further study was for the Minuteman system. This was called the Mark 12.

In July 1963, the Ballistic Missile Division issued a request to contractors for proposals for a MRV system to go on improved versions of the Minuteman missile. A two-part contract was awarded to General Electric in October. Part one was to research, design and flight-test the re-entry vehicle. Part two was for a nine-month study of a deployment mechanism for a multiple warhead configuration and for penetration aid and electronic counter-measures. The study was to include an examination of the feasibility and desirability of using various combinations of penetration aids and re-entry vehicles.

By the time the contract was issued to GE, the technical community was already aware of the central concept of a manoeuvrable bus. It had been conceived and suggested to the military in the period 1962–63. However, because of the desire to use the Mark 12 on the Minuteman II and technical uncertainties, the authorised design for the MRV warhead did not include independent, post-boost guidance. The Navy also had an exploratory design for a new missile (the Polaris B-3) in 1963, but that did not include MIRV either.

The crucial year of decision for MIRV was 1964. Delays in the Minuteman II schedules led to a decision to deploy the old Mark II re-entry vehicle rather than the experimental Mark 12 on which GE was labouring. This freed the Mark 12 development from the rigid production time-constraints of the Minuteman II programme. This in turn led to a decision to proceed with a Mark 12 system with a genuine MIRV capability together with a larger B-3 Navy missile (later the Poseidon C-3) which would also be MIRVed. At the end of the year the GE Mark 12 programme was reoriented to include a MIRV system. In January 1965 the guidance system subcontractor (North American Rockwell) was authorised to prepare plans for full-scale development of a Minuteman II post-boost control system, i.e. a MIRV bus.

The decision to develop fully the MIRVed missiles was made during 1965, as part of the five-year force plan. The go-ahead for the Poseidon missile programme was given towards the end of that year, and work on the Mark 12 continued, initially for the Minuteman II and then, after April 1966, for the more powerful Minuteman III missile. Production contracts for both the Minuteman III and the Poseidon missile were issued in 1968 and the missiles were first deployed in 1970 and 1971, respectively.

MIRV technology therefore emerged from a process that began with a clear decision by the Department of Defense to investigate responses to the potential deployment of ABM systems. This decision was made in the context of the enormous political impact of the launch of the first Soviet ICBM and Sputnik in 1957, and of Kruschev's famous boast that Russian ABM systems could 'hit a fly' in outer space.

However misplaced or premature American fears about Soviet missile interceptors might have been, it cannot be said that the development of ABM systems and penetration techniques proceeded as a result of quiet and uncoordinated investigations in weapons laboratories without stimulus or oversight from the Department of Defense.

However, the transition from decoys and MRVs to MIRV technology does appear to have occurred as a result of an evolutionary process with little political scrutiny. It brought together a number of different technologies developed in the field of propulsion and guidance systems, the launching of multiple satellites and ABM penetration devices.[60] A MIRV system could initially be developed by combining and extrapolating from these technologies purely on the initiative of the R & D laboratories. Unlike most new weapon systems, which require substantial funding and associated authorisation, the conceptual design of the MIRV development could be pursued under existing programmes that had already been authorised.[61]

Manifestly, MIRV was a technology whose time had come. In fact, Greenwood has identified no fewer than five 'independent' inventors of the MIRV concept: RAND, Aerospace Corporation, Lockheed, STL and North American Aviation. It was developed in advance of there being any specific strategic or mission requirement for it; although there had been strategic interest in multiple warheads to overwhelm potential ABM systems, there was no prior request for *independently targeted* warheads.

However, having recognised the key role played by R & D in independently developing the MIRV concept, it would be wrong to assume that its subsequent progress through to deployment was an uncomplicated story of the acceptance and development of the technology. From the time when the MIRV concept reached the offices of the services and the Department of Defense, it was subject to inter-service rivalry, bureaucratic interplay, commercial interests and political influences. Indeed, Greenwood's book is mainly concerned with tracing this complex process. The United States Air Force and Navy were initially resistant to the adoption of the MIRV concept. The Air Force preferred to build many more single Minuteman missiles, and the Navy feared that the Poseidon programme would divert resources away from the surface fleet.[62] It was not until Defense Secretary McNamara firmly limited the Air Force to 1,000 missiles in 1964 that it swung behind the MIRV programme. Likewise, the Navy only fully adopted MIRV when it became clear that the Poseidon submarine programme was going ahead and there was a danger that the Air Force might otherwise acquire a monopoly of the technology. Furthermore, there was clear pressure from Lockheed by 1963 for a defence programme to follow on from the Polaris A-3, which by then was coming to the end of its development cycle.

In one respect Greenwood's account appears to reinforce the relevance of the Zuckerman hypothesis to the adoption and full development of MIRV technology. Advocates of the MIRV system made strenuous efforts to devise strategic rationales to generate a mission requirement for their project. These rationales had to change as McNamara switched American nuclear policy away from counterforce strategies (as announced in his Ann Arbor speech in 1962) to the policy of mutual assured destruction (MAD). As Greenwood explains:

> In 1964 MIRV could be justified as a means of performing the counterforce mission, but by 1965 counterforce had taken second place to penetration as the major concern of the [Defense] Secretary and his staff. From then on, MIRV had to be sold as a penetration device.[63]

Superficially this could be taken to imply that counterforce strategies were essentially invented or discarded according to the requirements for selling the MIRV technology, which was being developed more or less for its own sake. However, such a conclusion would be wrong. Counterforce strategies were formulated independently of, and prior to, the adoption of the MIRV concept. The Defense Department could adopt MIRV technology as a means of substantially improving the counterforce capability of American forces, in line with existing operational strategy. It is now well-documented that MAD was adopted as American *declaratory* strategy in 1965 mainly as a means of limiting the budgetary requests of the military, and that counterforce doctrines remained at the heart of the country's *operational* strategy.[64] Thus MIRV remained consistent with operational strategy throughout — the contortions involved in some public justifications of the programme were mostly a reflection of the general problems caused when there are major differences between declared and operational doctrine.

The overall conclusion of Greenwood's analysis is that, in the case of MIRV, it was organisational and bureaucratic politics rather than the technologists which drove the decision-making process, in a set of interactions more complex than anything suggested by Zuckerman. 'As important as DDR & E and the technical community were', he writes,

> they could not have advanced MIRV beyond exploratory development on their own. Acceptance by the Secretary of Defense was essential for the initiation of full-scale development. The involvement of McNamara himself, at least as early as 1964, belies any claim that MIRV was just a product of the defense community. It was just as much a product of McNamara's policy preferences.[65]

The case of MX

The MX (Missile Experimental) programme began officially in June 1973 when the United States Air Force established an MX Office at Norton Air Force Base in San Bernadino, California. However, as Lauren Holland and Robert Hoover explain,[66] the Air Force and defence contractors had been exploring the concept and the technology of an advanced intercontinental ballistic missile (ICBM) since the mid 1960s. The MX Office at Norton brought together a number of existing, but previously separate, programmes on guidance, re-entry vehicles and warheads. During the mid 1960s neither the Department of Defense nor the Air Force gave high priority to a new ICBM to replace the Minuteman. But a number of interested groups and individuals within Strategic Air Command (SAC) and the defence industry pressed on with research for improving existing levels of performance:

> The absence of a pressing need for a new ICBM [in the 1960s] did not deprive this research [on a new ICBM] of a focus. Since the laboratories working on advanced ICBM technology already had helped develop earlier missiles, they simply concentrated on surpassing the 'effectiveness' of their previous designs. Effectiveness in missiles guidance systems is generally equated with precision in directing a missile to its target. Therefore, laboratory engineers tried to develop systems that would make a new ICBM far more accurate than existing ones. Similar efforts were made to improve missile propulsion. A crucial measure of effectiveness in propulsion systems is the amount of missile 'payload' they can

launch. By working on more efficient fuels, better engine designs, and related developments, engineers tried to increase the payload-lifting potential that a new ICBM might have [and thus the number of reentry vehicles and warheads the missile could transport].[67]

The impetus provided at this stage by the weapons scientists, stimulated in part at least by the technical challenge of the programme, is consistent with the Zuckerman hypothesis. The involvement from the beginning of interested groups within the Air Force, however, whose interest in developing war-fighting capabilities reinforced the research focus of the design laboratories of the weapons industry, suggests a rather more complex bureaucratic model than Zuckerman offers. It is also true that the 'improvements' being made in the missile's performance were fully in line with the strategic requirement to improve counterforce capability.

In the mid 1960s Strategic Air Command proposed to develop a much larger warhead for a proposed follow-on missile, with a warhead development programme called WS 120. The office of the Secretary of Defense and particularly Secretary McNamara himself, however, resisted this programme and eventually persuaded the Air Force to MIRV its existing Minutemen missiles, rather than to develop and deploy a new and larger missile. In the late 1960s, after McNamara's departure, the Air Force proposed a new research programme—WS 120A—to explore the idea of a MIRVed missile which was also larger than that of the Minuteman. This programme too encountered stiff resistance, this time from a more numerous opposition. Many members of the executive and the legislative branch believed that the development of a counterforce capability for the ICBM force would be destabilising: it could cause the Soviet Union to adopt a launch-on-warning attack posture. Potential opposition in the Congress on these strategic grounds led the office of the Secretary of Defense eventually to reject requests for further funding for the WS 120A programme. As Holland and Hoover comment:

> It is interesting to note that political officials outside the Pentagon and executive branch were disturbing the inner layer of this procurement decision almost from the very beginning; and the strategic issue motivated interests outside the Pentagon and executive branch to attempt to influence the procurement process.[68]

The Zuckerman hypothesis does not give rise to optimism about the chances of political action being taken to curtail potentially destabilising research programmes in this way. The intervention of Congress during this early stage of weapons research on strategic grounds is also contrary to the more complex bureaucratic model. This model sees congressional interest in procurement matters as coming at a later stage and being motivated by 'pork barrel' (that is, the attempt by con-

congressmen to secure weapons contracts for their own districts) rather than by strategic considerations.

In 1971 the Air Force, intent on developing weapons systems that would allow Strategic Air Command to improve on the counterforce capability of its nuclear forces, filed an official required operational capability (ROC) request to allow it to develop a new, larger, more accurate MIRVed missile to be the follow-on to the Minuteman programme. The proposal was to integrate existing programmes on advanced ICBMs, including research and development on newer, heavier missiles, with increased warhead kilotonnage and improved accuracy. This proposal, the basis for what eventually became the MX design, merged hundreds of disparate decisions made by different groups during the research and development of advanced ICBM technology.[69] During this process it seems that it was the Air Force who provided the overall specification. The role of scientists was that of generating ideas about how best to meet the objectives they were given:

> The key organisational actors affecting design decisions for an advanced ICBM system during this period were SAC and the design laboratories of major defense corporations. As the future proprietor of the new ICBM, SAC wanted and got the performance characteristics that matched its mission's objectives. The design laboratories furnished the ideas for how to improve the accuracy and throw-weight characteristics of the new missiles in comparison to the previous Minuteman program.[70]

The MX programme was formalised in 1973 and the Air Force duly opened its MX office in California. Between 1973 and 1976 the Air Force and the Department of Defense began research, development, and testing of the MX missile and several basing schemes were evaluated. During this, the engineering phase of procurement, the Department of Defense first asked Congress to authorise funds for MX deployment. Four organisations within the Department of Defense—Ballistic Missile Division (USAF), Strategic Air Command, Director of the Department of Defense Research & Engineering (DDR & E), and the office of the Secretary of Defense—played a key part in influencing MX decision making during this period, along with the National Security Council, and the prime contractor for research and development, TRW Corporation. The National Security Council was particularly concerned with the MX since it had strategic implications for the current SALT negotiations. The intervention of Secretaries of Defense James Schlesinger and Donald Rumsfeld—who saw MX as providing the hard-target potential necessary for a damage limitation strategy—were critical for the survival of the programme in 1975 and 1976. The strategic implications were also critically significant to Congress and had important effects on MX decision making.

It is clear that even during the early stages of MX procurement, focusing on design, research, development and testing, the role of the weapons scientists was nothing like as central as that suggested by the Zuckerman hypothesis. Certainly in the earliest stages the ingenuity and resourcefulness of industrial engineering groups within the defence community interacted with Air Force doctrine and interests in a way compatible with a more complex bureaucratic model. But by the time of the engineering phase of procurement, the strategic and foreign policy implications, of concern both to the executive branch and to Congress, were important factors in shaping the character of the MX. From that point onwards, as the question of basing modes for the MX became increasingly prominent, a whole range of other groups, many of them previously unconcerned with defence issues as such, became involved:

> These new groups brought to the procurement process a set of previously overlooked concerns: environmental impact, socio-economic costs, fiscal and budgetary demands, and strategic and foreign policy implications. In addition, these groups compelled state and local governments within the region targeted for MX deployment to become involved intimately in the decision-making process, thus activating unused governmental arenas.[71]

By the creative use of legislation such as the National Environmental Protection Act of 1969, the Federal Land Policy and Management Act of 1976 and the Freedom of Information Act of 1965, these groups had a critical impact on the later stages of MX decision making.

The case of MX does not, of course, of itself disprove either the Zuckerman hypothesis, or the more complex bureaucratic model of decision making. As Holland and Hoover acknowledge, procurement programmes that require relatively few resources, have minor environmental or social costs, do not have a nuclear dimension nor raise strategic questions may well fit bureaucratic models of decision making better than the MX case does. What they do argue compellingly is that the MX programme was a manifestation of significant and durable changes in the pattern of procurement decision making for major weapons systems in the post-Vietnam/Watergate era. If this is the case, not only is the Zuckerman hypothesis only a slight guide in the case of the MX; even the more sophisticated bureaucratic model of decision making itself could be inadequate in explaining major examples of weapons procurement in the United States in the foreseeable future.

Strategic cruise missiles

As a land-based system with massive environmental implications, the MX missile was perhaps an untypical case. The development of long-range cruise missiles appears at first sight to be a more promising exemplar of the Zuckerman hypothesis. Like the MIRV concept described above, the strategic cruise missile emerged from a series of uncoordinated innovations in separate programmes.[72] During the 1960s small, efficient, turbofan jet engines were developed that would allow a new missile to achieve a range of about 1,500 miles while still remaining very small. Advances in warhead design allowed compact and efficient nuclear or conventional warheads to be produced. New designs and strong lightweight materials allowed the production of a light, rigid airframe with a very small radar image. Most importantly, advances in micro-electronics, mini-computers and map-making techniques made the terrain contour matching (TERCOM) guidance system—which was first patented in 1958—practicable, allowing accuracies of a few tens of metres to be achieved.

These technologies were first brought together in a programme in the mid 1960s to develop decoy cruise missiles for B-52 bombers as an aid for penetrating Soviet air defences. Scientists within the DDR & E became intrigued by the technical possibilities, and began to develop the strategic cruise missile (SCM) concept. By 1969 the Air Force was becoming alarmed that their decoy was developing into a long-range cruise missile that could undermine the role of the penetrating, piloted bomber. They tried to insist that the development be confined to the defined mission requirement—a short-ranged decoy.

Meanwhile, the United States Navy was engaged in projects to develop short-range, anti-ship cruise missiles. After the development of the 70-mile range Harpoon missile, the Navy started an advanced cruise missile programme early in 1971, which was intended for the same purpose, but with a longer range (300 miles). Once again, under pressure from the DDR & E, this programme began to explore the possibility of producing an SCM. In both the Air Force and the Navy programme there followed a period of several years during which the services resisted the adoption of the SCM concept, against pressures exerted by a 'small group of technically oriented civilians' in the office of the Secretary of Defense, particularly the research and development division (DDR & E).[73] The Navy's aircraft carrier supporters saw the long-range cruise missile as a potential threat to their aircraft and the submarine departments warned that it could divert funds away from Trident. The Pentagon cancelled the USAF's decoy project in July 1973 and forced the Air Force to embark on a strategic air-launched cruise missile (ALCM) project. For more than a year after that the Air Force still tried to protect its perceived institutional interest by trying to restrict the ALCM's size, and therefore its range, by insisting that the new missile should fit existing bomber missile racks.

It can therefore be seen that the military R & D institutions played a key role in developing the concept of an SCM and in arguing for its adoption in the face of serious resistance. However, it should be noted that the DDR & E is not a weapons laboratory but rather a division of the Defense Department that is specifically tasked to identify and investigate promising new technologies with potential military applications. More importantly, in this context, the DDR & E is unlikely to have been able to sustain the strategic cruise missile programme without significant political support.

In July 1972 Secretary of Defense Melvin Laird proposed the development of a submarine-launched cruise missile to the Senate Armed Services Committee, as part of the bargaining process involved in securing the support of the Joint Chiefs of Staff and Senate 'hawks' for the SALT I Treaty. The rationales provided by Laird for the development of such a missile were fairly unconvincing. It was presented as a response to the short-range cruise missiles on Soviet submarines and also as a 'bargaining chip' for arms control negotiations to encourage the Soviet Union to agree to restrict its own cruise missile forces. Later in the same hearings Laird said, 'the development of the SLCM is necessary to *assure availability* of *future* US *options* for *additional* US strength *if needed*'.[74] This became known as the 'Laird hexahedge' and indicates that the programme was driven at that stage by institutional and political pressures rather than by clear strategic requirements. However, the historical accounts indicate that these pressures were generated at least as much by the politics surrounding the SALT process as by the R & D institutions.

The SALT process came to the rescue of the foundering cruise missile programme again in 1974 and 1975. Henry Kissinger was looking for 'marginal' long-range systems that could be bargained away in the forthcoming SALT II negotiations, and he exerted strong pressure on the Navy and the Air Force to continue with the cruise missile project.

By 1976 the development programme was showing impressive results, and the SCM began to acquire wider support from those who saw it as a cheap and versatile weapon. However, as Huisken states:

> The strategic cruise missile did not acquire a strong doctrine-related rationale until mid-1977 when it displaced the B-1 [bomber]. At that time the view was put forward that the cruise missile would effectively counter the emerging preponderance in hard-target kill probability [i.e. counterforce capability].[75]

In June 1977, the Carter administration cancelled the B-1 bomber programme and substituted a plan to produce some 3,418 ALCMs for the B-52 fleet of bombers. The SCM programme was then firmly established as a major weapons project. Major companies such as Boeing and General Dynamics became heavily committed to the programme and the services revised their hostile attitude.

The history of the SCM programme certainly shows that SCMs

evolved without a well-defined conception of why they were needed. It also demonstrates that the military R & D establishments played a key role in the early stages. However, as in the case of MIRV technology, any model which focuses on the role of scientists would grossly distort the evolution of the programme in which bureaucratic and commercial interests and international and domestic political processes played an equal if not greater role.

Chevaline

The Zuckerman hypothesis has not proved a particularly powerful explanatory tool for the American weapon systems considered so far. Scientific activities beyond the limits authorised by political leaders and the defence bureaucracies have been significant, as has been the influence of military R & D institutions. But they have been only one factor amongst many (often far from the most important) in determining the future of the weapon system involved, particularly after it has moved from the initial, conceptual stages through adoption to full development and deployment. However, the hypothesis might explain better the processes of nuclear weapons development in Britain, with its much higher levels of secrecy than the United States has. Zuckerman's perspective on these issues was probably most influenced by his experience within the British Ministry of Defence, ultimately as Chief Scientific Adviser between 1960 and 1966 and similarly to the Prime Minister for a further seven years. We have already said that he identified the British Chevaline project as a prime example to support his hypothesis.

Like the MIRV programme, Chevaline began in the context of concerns about Soviet ballistic missile defences and their potential implications for the effectiveness of the British Polaris missile force. The Polaris A-3 missile, which the United Kingdom had bought from the United States, already had some potential for overwhelming endo-atmospheric ABM systems similar to the American Nike–Zeus. The three re-entry vehicles on the A-3 were designed to separate sufficiently so that 'even when fired from minimum range, one Zeus-type interceptor could not destroy all three'.[76] However, such measures might not be effective against an exo-atmospheric ABM, such as the Soviet Galosh interceptor, which was apparently first observed during the Moscow parade on 7 November 1964.[77] In principle one such interceptor could disable all three incoming warheads in space where they would be still relatively close together.

When the Galosh ABM system was deployed around Moscow in the late 1960s, there were suggestions that Britain should upgrade the 'front end' of its Polaris missiles to increase the probability that they could penetrate any Soviet defences. Even before *HMS Resolution*, the first British Polaris submarine, went on patrol in June 1968, the atomic Weapons Research Establishment (AWRE) at Aldermaston started seriously to examine the potential implications of the Galosh system.[78] Peter Jones, head of AWRE Warhead Electronic Firing Systems Division in 1968, testified just before he was promoted to Director of AWRE in 1982 that 'By that time [1967–68] we realized that [Galosh] was very significant in terms of the effect on Polaris and that we would have to take any changes we made to Polaris very seriously indeed'.[79] AWRE alerted MoD civil servants and the inner Cabinet to the issue and Prime Minister Harold Wilson authorised it to undertake studies into potential countermeasures.

Peter Jones, who rose to the highest post in AWRE through his association with Chevaline, directed these studies and in 1970 the new Conservative government authorised a feasibility study on a system which would involve multiple warheads, decoys and hardened warheads to withstand radiation.[80] The design concept adopted was American in origin where it was known as Antelope. It had been superseded by MIRV technology in the United States, but in Britain £7.5 million was allocated in 1970 to develop the concept and define a project to upgrade the penetrative ability of Polaris (codenamed Chevaline). This project was completed in 1972 when it was estimated that a five-year development programme followed by production would cost £175 million (autumn 1972 prices).

Between 1972 and 1974 the option to proceed with Chevaline development was kept open on the basis of interim funding for successive three- and six-month periods. Meanwhile, other options were considered.[81] The decision to be made was eventually between developing Chevaline and buying the American MIRVed Poseidon missile.[82] Poseidon had the advantage that it was already fully developed and therefore less prone to cost-escalation. It was also a more militarily capable system and (initially) preferred by the Royal Navy.

However, a number of political factors counted against Poseidon. Prime Minister Heath was given the firm impression during a visit to Washington that 'while, if asked, the President would agree to a transfer of Poseidon, he preferred not to be asked'.[83] Although the United States had previously supplied Britain with technical information on Poseidon and had indicated that it was available for sale to the United Kingdom, the political climate had changed.[84] Congress, which was controlled by the opposition Democratic Party, was intent on asserting more control on defence issues (partly as a result of the unpopular Vietnam War) and many congressional representatives were opposed to MIRV technology and to the transfer of defence technology to other nations.

The purchase of Poseidon was not politically attractive to the British Government either. The Labour opposition would probably have opposed it, thus destroying the bipartisan consensus on nuclear weapons policy established since the Wilson Government accepted Polaris.[85] It could have rekindled public opposition to the British nuclear forces. The British Government must also have been sensitive to the possible impact that the transfer of MIRVed missiles could have on the SALT II negotiations: it would have reinforced demands that British forces be included in the talks.

The argument in favour of Chevaline was clinched when the cost estimate of the Poseidon option was reassessed in late 1973, raising it from £250 million to £500 million.[86] In contrast, the cost of Chevaline was estimated at £255 million. The Heath Government opted for the Chevaline project in January 1974. Then in a surprise election victory, Labour was returned to power one month later. A small Cabinet subcommittee, consisting of Wilson, Healey, Jenkins, Callaghan and Mason, agreed to proceed with Chevaline. This decision was confirmed in September 1975, but the full Cabinet was only briefly informed of it. It was kept secret from Parliament and the public until 1980.

There is good evidence that AWRE had a significant influence over the development of Chevaline. AWRE personnel raised the issue of counter-measures to Galosh within the MoD and stimulated the authorisation of initial studies. Managers and decision makers were also very dependent on advice from Aldermaston. Until 1977 Hunting Engineering Ltd were the co-ordinating design authority for the programme, and they lacked adequate technical expertise in many relevant areas and had to rely on advice from AWRE.

The Government itself exerted only loose and sporadic control over the programme. It was a potential political embarrassment, particularly for the Labour Government. Ministers apparently preferred to avoid close engagement with the programme. For years the programme was only drip-fed, with funds authorised for only a few months ahead. C. J. Carey, the then Treasury Officer of Accounts, later recalled that 'Ministers were still not prepared to authorise funds for more than a year at a time' in 1974–76.[87]

However, although Ministers were not enthusiastic about Chevaline, they did not want to be accused of failing to maintain an effective deterrent. So at each stage they accepted the recommendations of AWRE, MoD civil servants and the commercial subcontractors and allowed the programme to go one step further.

One indication that the Chevaline programme was 'supply-driven' rather than 'demand-led', is that the signing of the ABM Treaty in 1972 apparently had little impact on British Government considerations. This treaty strictly constrained Soviet ABM systems, limiting the

Soviet Union to one ballistic missile defence system around Moscow. It thereby removed much of the rationale for Chevaline, especially as the single existing Galosh system was widely judged to be of very limited effectiveness. Air Vice-Marshal Stuart Menaul expressed a typical expert view when he said:

> Examination of the types of radar in the Galosh system and the state of Soviet computer technology, and even in the guidance systems employed in the interceptor missiles, would indicate that the efficiency of the Galosh system in shooting down ballistic missiles would probably be less than twenty per cent.[88]

By the time firm management and control of the programme was finally imposed in 1977, in response to escalating costs, and power shifted away from AWRE, so much had been invested in the programme that it was decided to complete the Chevaline project even though ministers acknowledged that the strategic requirement for the system was very slight. David Owen, Foreign Secretary from 1977 to 1979, has since revealed that at Cabinet meetings in 1977 and 1978, when the cancellation of Chevaline was last considered, it was recognised that 'most of the Chevaline money had already been spent or was already contracted for' and that 'the only credible arguments for continuing were the political and not the military aspects of our deterrence strategy'.[89] Overall, Chevaline was officially judged to have cost £1,000 million and unofficially it has been calculated to have cost as much as £1,670 million (September 1980 prices).[90]

Overall then, the Zuckerman hypothesis does appear to find some support from the case of Chevaline. AWRE had a strong influence in stimulating and shaping the Chevaline programme and had a clear institutional interest in advancing the project. It provided the institution with technically challenging work and increased resources. It was a means by which AWRE could enhance its status amongst its American counterparts and perhaps hope that the Anglo-American nuclear information exchange did not always operate in one direction. It allowed AWRE to maintain and strengthen its design teams. Indeed, the last two advantages were frequently cited in arguments by AWRE and MoD officials and other supporters of Chevaline. Politicians were often presented with a choice of either supporting Chevaline or allowing Aldermaston's capacity to build new nuclear weapons to atrophy and the Anglo-American nuclear relationship to unravel.

However, it is important to recognise the importance of politics in the development of Chevaline. As we have seen, the reasons why Chevaline rather than Poseidon was adopted were primarily political, although AWRE and MoD civil servants were influential through their role in producing cost estimates for the two projects, estimates that turned out to be very unrealistic. The reasons why government allowed a loose managerial control over the project in the critical years before 1977 were apparently political; it was only in that context that AWRE was able to retain so much control over the project. And the main reasons why ministers always decided to allow the project to continue, despite their manifest lack of enthusiasm and the weakness of the strategic rationales, were surely also political. Arguments about maintaining Aldermaston's design teams and the respect of the American nuclear establishment were influential amongst policy makers only because they had accepted that Britain should try to retain its credibility as a major nuclear weapons state, and that this implied that Britain must be seen to maintain both modern nuclear forces and a technical infrastructure capable of independent R & D on nuclear weapons.

It is widely recognised that the independent nuclear weapons policy pursued by Britain is primarily driven by political considerations.[91] British governments have long supported nuclear programmes for which the strategic arguments are weak, and the fact that this occurred in the particular case of Chevaline should not come as a surprise. In summary, AWRE did play a significant, independent role in promoting and sustaining the Chevaline project, but political factors were probably more important. Furthermore AWRE, and also a number of commercial subcontractors, would not have been able to exert such strong influence were it not for the particularly loose managerial control, which was again primarily a result of politics. The Zuckerman hypothesis usefully alerts us to the role of AWRE, and also to the fact that decision makers seemed to be over-reliant on it for advice, but it also tends to blind us to the importance of policy and politics. In doing so it is so partial as to be seriously misleading.

Conclusion

We have examined a number of cases in which the Zuckerman hypothesis seems most likely to apply. In some cases the hypothesis seems to have little relevance. In others it rightly directs attention to the role of uncoordinated technological development and to the influence of military R & D institutions, but it also distracts attention from other equally, if not more, important factors. The hypothesis therefore provides a poor framework with which to understand the way in which these new weapons systems were developed.

It is possible that there are some instances where the Zuckerman hypothesis is correct in identifying the autonomous activities of scientists and scientific institutions as the dominant driving force for initiating and shaping new weapons development. However, the SDI programme is certainly not such an instance and neither is Trident D5, the B-1B bomber, the Tornado nor Midgetman.[92] The development of 'third generation' nuclear warheads is perhaps one of the most likely candidates.[93] The Los Alamos and Lawrence Livermore nuclear weapons laboratories in the United States have taken a key role in promoting and developing these systems. Yet even here strategic and political considerations have played an important role, and in most cases, the development of these new designs is still at an early stage. In those instances where the warhead is at a more advanced stage, such as in enhanced radiation (neutron) bombs and earth-penetrating warheads, strategy, politics, and bureaucratic interests have clearly played a critical role in shaping their development and deployment.

R & D programmes frequently 'over-achieve' or stimulate interest in new military possibilities, but they do so in a context shaped by broader political, strategic, bureaucratic and financial factors. Similarly, military R & D institutions are frequently influential, particularly at the earliest stages of the development programme, but they comprise only one factor amongst several in the process, and rarely the most influential.

Nevertheless, it is remarkable how much vitality the Zuckerman hypothesis appears to have, particularly amongst those who do not study the details of how weapons have actually been developed and procured. The hypothesis has operated as an important explanatory myth, with the effect that attention has been diverted from the role of politicians and other groups and processes. As the H-bomb case study showed, Truman greatly exaggerated the development work that had already been carried out on the bomb, when he announced his decision to proceed with the programme. Could it be that he found it rather convenient to imply that R & D had gone much further than in fact it had in order to make it seem that his policy decision was only part of a natural continuum and so reduce the political opposition to it?

Focusing on the autonomous role of R & D makes the arms race appear more inevitable—an inescapable feature of progress or of the modern world. It absolves the politician and others of responsibility, while identifying the guilty party as a limited, mysterious group with professional, vested interests. By drawing on the prestige and mystique of science, the myth demoralises or diverts potential critics.

It is probably natural for scientists, familiar as they are with just one aspect of the development of a weapons programme, to emphasise the role of the R & D institutions. It may be significant that, of those concerned in weapons procurement who have subsequently written about it, it is scientists themselves—notably Zuckerman and York— who have laid most stress on their role. General Groves, military commander in chief of the Manhattan project and one of the few such military personnel to write about his experience, gives a very different picture of the role of scientists—as useful servants who need strict supervision to prevent their indiscipline and scientific curiosity from leading them to toy with intriguing but militarily marginal ideas.[94]

Above all, the Zuckerman hypothesis is a beguiling mythology because it symbolises a view that the weapons development process is out of rational control and is supply-driven rather than demand-led. Our brief case studies showed that this view contains a good deal of truth. Weapons systems tend to emerge as a result of institutional and social processes, not to fulfil clear strategic demands. However, a supply-model focusing on institutional interest and interactions must be much more complex than the Zuckerman hypothesis before it can even begin to explain weapons development programmes. Even complicated supply-models omit essential factors. Politics cannot be left out of account and, indeed, strategic demand often plays an important role.

Perhaps the reasons for the wide appeal of Zuckerman's hypothesis are more interesting than the substance of the hypothesis itself. Instead of dealing with the complexity of the process, it satisfies our urge to find a villain, or a single cause for the arms race, and can either offer the hope of seductively easy but illusory solutions or absolve most of us of responsibility.

Notes

1. Roger Williams, 'British scientists and the bomb: the decisions of 1980', *Government and Opposition*, Vol. 16, No. 3, Summer 1981, p. 274.
2. Lord Zuckerman, 'The deterrent illusion', published in *The Times*, 21 January 1980.
3. Solly Zuckerman, *Nuclear Illusion and Reality*, Collins, London, 1982, p. 101.
4. Ibid., p. 106.
5. Ibid., p. 107. However, the only evidence supporting this particular assertion consists of a pair of reports in the *Sunday Times* and the *Guardian* newspapers.
6. Zuckerman, see note 3, p. 108.
7. President Eisenhower's farewell address, 17 January 1961, reprinted in *Air Force Magazine*, October 1983.
8. Herbert York, *Race to Oblivion: a Participant's View of the Arms Race*, Simon & Schuster, 1970.
9. Herbert F. York, 'The origins of MIRV', in David Carlton and Carlo Schaerf (eds.), *The Dynamics of the Arms Race*, Croom Helm, 1975; Graham T. Allison and Frederic A. Morris, 'Armaments and arms control: exploring the determinants of military weapons', in *Arms, Defense Policy and Arms Control*, Franklin A. Long and George W. Rathjens (eds.), Norton, 1976.
10. Lord Mountbatten, Strasbourg address of 11 May 1979, published in *Apocalypse Now?*, Spokesman, Nottingham, for the Atlantic Peace Foundation and World Disarmament Campaign, 1980.
11. Quoted in Council for Science and Society, *UK Military R & D*, Oxford University Press, 1986, p. 50.
12. Ralph Lapp, *Arms Beyond Doubt: the Tyranny of Weapons Technology*, Cowles Publishing Company, 1971.
13. Dietrich Schoeer, *Science, Technology and the Nuclear Arms Race*, John Wiley, 1984.
14. Lauren H. Holland and Robert A. Hoover, *The MX Decision: a New Direction in US*

Weapons Procurement Policy?, Westview, Boulder, 1985, p. 12. The organisational process view derives from Graham Allison, *Essence of Decision: Explaining the Cuban Missile Crisis*, Little, Brown, Boston 1971. See also A. W. Marshall, *Bureaucratic Behavior and the Strategic Arms Competition*, California Arms Control and Foreign Policy Seminar no. 4, October 1971.
15. Thomas A. Brown, *What is an Arms Race?*, California Arms Control and Foreign Policy Seminar no. 35, October 1973.
16. Ibid.
17. Mary Acland-Hood, 'Military research and development expenditure', in *SIPRI Yearbook 1986*, London, p. 299.
18. Harvey Brooks, 'The military innovation system and the qualitative arms race', in Long and Rathjens, see note 9.
19. York, see note 8, p. 179.
20. Rathjens interviewed in 1985 for the second television programme of the Open University course, Nuclear Weapons: Enquiry, Analysis and Debate.
21. *SIPRI Yearbook 1983*, London, p. 214.
22. Martin Binkin, Herschel Kantler and Rolf H. Clark, *Shaping the Defense Civilian Workforce*, Brookings Institution, Washington, 1978, p. 8.
23. *UK Military R & D*, see note 11, p. 9.
24. *Statement on the Defense Estimates 1987*, Cm 101-II, HMSO, 1987.
25. Eisenhower, see note 7.
26. Brooks, see note 18, p. 14.
27. Ibid., p. 87.
28. Holland and Hoover, see note 14, p. 14.
29. Mary Kaldor, 'The weapons succession process', *World Politics*, July 1986, p. 577.
30. Warner R. Schilling, 'Scientists, foreign policy and politics', in Robert Gilpin and Christopher Wright (eds.), *Scientists and National Policy Making*, Columbia University Press, 1964, pp. 159–60.
31. York, see note 8, p. 157.
32. Albert Wohlstetter, 'Strategy and the natural scientists', in Gilpin and Wright, see note 30, pp. 197–8.
33. York, see note 8, p. 179.
34. Brooks, see note 18, pp. 90–1.
35. See Robert S. de Ropp, *The New Prometheans: Creative and Destructive Forces in Modern Science*, Cape, London, 1972.
36. Edward Teller with Allen Brown, *The Legacy of Hiroshima*, Doubleday, 1962.
37. York, see note 8, pp. 32–3.
38. Eisenhower, see note 7.
39. W. Broad, *Star Warriors — the Weaponry of Space: Reagan's Young Scientists*, Faber & Faber, London, 1985.
40. Brooks, see note 18, p. 76.
41. Quoted in *The Independent*, London, 31 March 1987.
42. Brooks, see note 18, p. 76.
43. *Public Papers of the Presidents of the United States, Harry S. Truman*, U.S. Government Printing Office, Washington D.C., 1961–66, p. 138.
44. *The Journals of David E. Lilienthal*, Harper & Row, New York, vol. II.
45. Teller and Brown, see note 36, p. 45.
46. Testimony of Norris Bradbury in *In the Matter of J. Robert Oppenheimer*, U.S. Government Printing Office, Washington D.C., 1954, p. 485.
47. Jonathan B. Stein, *From H-Bomb to Star Wars: the Politics of Strategic Decision Making*, Lexington Books, 1985, p. 6.
48. Quoted ibid., from transcript of a Teller lecture, p. 6.

49. Ibid., p. 7.
50. Testimony of Isidor Rabi in *In the Matter of J. Robert Oppenheimer*, see note 46.
51. Stein, see note 47, p. 49.
52. Lewis Strauss, an AEC Commissioner, reflected, 'It is sobering to speculate on the course of events had there been no monitoring system in operation in 1949. Russian success in that summer would have been unknown to us. In consequence, we would have made no attempt to develop a thermonuclear weapon. It was our positive knowledge of Russian attainment of fission bomb capabilities which generated the recommendation to develop a qualitatively superior weapon—this to maintain our military superiority'. Lewis Strauss, *Men and Decisions*, Doubleday, New York, 1962, p. 201.
53. Stein, see note 47, p. 23.
54. Ibid., p. 19.
55. Ibid., p. 2.
55. Ibid., p. 27.
56. Herbert York, *The Origins of MIRV*, SIPRI Research Report no. 9, 1973, pp. 8–14; Allison and Morris, see note 9.
57. Ted Greenwood, *Making the MIRV: a Study of Defense Decision Making*, Ballinger, 1975.
58. York, see note 9, p. 35.
59. Greenwood, see note 57.
60. York, see note 56, pp. 8–14.
61. York, see note 56, p. 22; see also Greenwood, see note 57, for an extended account of this.
62. S. McLean (ed.), *How Nuclear Weapons Decisions Are Made*, Oxford Research Group, Macmillan, London, 1986, pp. 79–80.
63. Greenwood, see note 57, Chapter 3.
64. D. Ball, 'The development of the SIOP, 1960–1983', in D. Ball and J. Richelson (eds.), *Strategic Nuclear Targeting*, Cornell University Press, Ithaca, 1980, pp. 57–83; D. Ball, 'Targeting for strategic deterrence', *Adelphi Papers*, no. 185, International Institute for Strategic Studies, London, 1983.
65. Greenwood, see note 57, p. 80.
66. Holland and Hoover, see note 14, p. 124.
67. Paul Stockton, quoted in Holland and Hoover, ibid., p. 125.
68. Ibid.
69. Ibid.
70. Ibid., pp. 126–7.
71. Ibid., pp. 3–4.
72. R. Huisken, *The Origins of the Strategic Cruise Missile*, Praeger, 1981; and R. Betts (ed.), *Cruise Missiles: Technology, Strategy, Policy*, Brookings Institution, Washington, 1981.
73. Huisken, see note 72, p. 187.
74. Quote by K. Tsipis, 'Long Range Cruise Missiles', in *Armaments and Disarmament: the SIPRI Yearbook 1975*, MIT Press, Mass., USA, 1975, p. 327.
75. Huisken, see note 72, p. 160.
76. Greenwood, see note 57, p. 161.
77. G. Spinardi, *The Chevaline Programme: a Case Study of the Social Control of Weapons Technology*, unpublished, University of Edinburgh, 1985, p. 4.
78. House of Commons Committee of Public Accounts, *Ministry of Defence: Chevaline Improvement to the Polaris Missile System*, 9th Report, HC Paper 269, 1981–82, para. 130.
79. Quoted in H. Miall, *Nuclear Weapons: Who's in Charge?*, Macmillan, 1987, p. 14.

80. House of Commons Committee of Public Accounts, see note 78, p. vi.
81. Ibid., MoD memorandum.
82. See, for example, historical account in Lawrence Freedman, *The British Nuclear Deterrent*, Macmillan, 1980, pp. 46–51.
83. P. Malone, *The British Nuclear Deterrent*, 1984, Croom Helm, p. 68.
84. Spinardi, see note 77, p. 12.
85. S. McLean (ed.), see note 62, p. 145.
86. House of Commons Committee of Public Accounts, see note 78, MoD memorandum, para. 5.
87. *Sunday Times*, 10 February 1980, quoted in Spinardi, see note 77, p. 15.
88. *The Times*, 6 July 1981, quoted in McLean, see note 62, p. 147.
89. Quoted in Zuckerman, see note 3, p. 107.
90. M. Chalmers, *Paying for Defence*, Pluto Press, 1985, p. 185.
91. For instance, Freedman, see note 82.
92. R. Bulkeley and G. Spinardi, *Space Weapons: Deterrence or Delusion*, Polity Press, 1985; G. Werken, 'The earthly origins of Star Wars', *Bulletin of the Atomic Scientists*, October 1987, pp. 20–8; D. Carlton, see chapter 2 *supra*.
93. For a brief description of these warheads see K. Tsipis, 'Third-generation nuclear warheads', *SIPRI Yearbook 1985*, Taylor & Francis, 1985.
94. L. R. Groves, *Now It Can Be Told: the Story of the Manhattan Project*, Deutsch, 1963.

PART II

Introduction
Civil–Military Interactions: Fragmentation, Distortion and Policy Making

The subjects explored in the previous chapters illuminated aspects of the relationships between science, scientific advice and the defence policy-making process. They illustrated the fact that the production of science and scientific studies relevant to defence is normally shaped by the broader political, social and economic processes and interests. The prestige of science, with its reputation for reliability and political neutrality, means that scientific reports enter the policy-making process as potentially valuable weapons for promoting certain policies and discrediting others. In this process the scientific community loses control of its product as qualifications are ignored and uncertainties are either exaggerated or neglected. Interest groups use scientific studies in order to pursue their objectives, and reports tend to be adopted or shelved according to political convenience rather than objective validity. The results of the policy-making process will, no doubt, have been affected by the original scientific studies but in unpredictable ways, and in ways that might surprise or offend the scientists involved. More than this, at every stage the overall policy-making process affects the environment in which continuing scientific work is being pursued.

These processes were particularly well illustrated in the first two chapters. Chapter 2, on the Strategic Defense Initiative, and chapter 3, on the Zuckerman hypothesis, also emphasised the important role of myths. For example, the myth of science as the almost magically potent servant of desired defence policies—an idea created in the first instance by the Manhattan project which produced the atomic bomb—appeared alive and well in President Reagan's television address of March 1983 on the SDI. The inverse myth—of science and scientists as the all-powerful and malevolent force behind an arms race which is out of political control—lies at the heart of the Zuckerman hypothesis. These mythologies continue to command a surface credibility for very good reasons: summoning up the wonders of modern science to solve the deep-seated fears of the nuclear age is a powerful political weapon.

Nor is this just mythology: to invest heavily in any science-based defence programme far ahead of the current state of the art is bound to produce major new military developments, although it is impossible to foresee exactly what these will be. In other circumstances it may be advantageous for political leaders to subscribe to a Zuckerman-type hypothesis, which transfers the blame for escalating levels of armaments from their shoulders to those of the weapons scientists in the laboratories. Moreover, not all weapons scientists are necessarily reluctant to have the importance of their role magnified in this way. Concerned citizens, baffled by the complexities underlying rising weapons levels, may seize with relief on an interpretation which confines the blame to a defined and limited set of individuals and thus reduces the complexities to an easily comprehensible set of ideas.

Whatever the role of science, bureaucracies, strategy or political ambitions in the defence policy-making process, decisions taken have extensive, and often poorly understood, ramifications, extending far into the civil sector. In Part II we widen our focus from the defence sector alone to the way that interactions between the military and the civil sector are managed, or not managed, as is often the case, concentrating for the most part on the United Kingdom whereas Part I concentrated on the United States.

Modern advanced technology characteristically underpins simultaneous military and civil applications, creating complex linkages across traditional boundaries. Nuclear energy and electronics, the subjects of chapters 4 and 5, are prime examples of such phenomena. The problems arising from such linkages are diverse and complex. Developments in one area, for example, will inevitably be competing in some sense for investment or for the services of skilled scientific personnel with developments in another. Policy decisions about the relative benefits of investments in each area are immensely complex, raising as they do technical questions in economics, as well as in security. Such decisions should in theory rest on an overall assessment of both military and civil areas: in practice, however, ministerial and bureaucratic divisions almost always exacerbate partiality and encourage adversarial rather than holistic approaches to these questions.

Ideally, policies should rest on as dispassionate an assessment of the available economic and political evidence as can be mustered. In practice the findings of the relevant social sciences are even more likely than the more prestigious physical sciences to be taken up, distorted or ignored by policy makers, according to their ideological preferences. Ideally, as much information about military/civil interactions as is compatible with the strictly defined requirements of national security should be in the public domain, as a resource for policy-relevant research. In practice it may be convenient to let secrecy deriving from the military area spill over into the civil area, to thwart investigation and accountability, and to cover up questionable management decisions. Since the military sector is, especially in the nuclear case, politically controversial, both in terms of domestic politics and of the international non-proliferation regime, governments may be embarrassed by open discussion of the nature of military/civil links. It may be more convenient to allow myths and misunderstandings to develop and remain uncorrected, or even, if driven to extremes, to issue directly misleading statements, which only laborious delving in the remoter public archives can contradict. In such circumstances the distorting effect of decisions taken in one area—and in major Western states the high prestige area is usually the dominant one—or the other can only be guessed at.

David Lowry's chapter, 'Nuclear weapons and nuclear power: bias and mythology in the making of the British Magnox nuclear reactor programme', exposes some of the myths which have been created, or allowed to develop, around the British military and civilian nuclear programmes. In the process it suggests some possible roles for secrecy, privileged knowledge, bureaucratic politics and geostrategic factors in the presentation and implementation of the British civil nuclear programme. It also illustrates the limitations of Parliament in overseeing a programme which is associated with powerful institutional interests, which at one and the same time can claim no connection with military programmes and yet shelter behind claims of national security and privileged knowledge, whenever convenient, and which can exploit the political appeal of high technology, big science and 'progress'. Critics have often been obliged to fight their case on scientific and technical issues where their relative lack of resources and hands-on experience works against them.

Myths of the separation between civil and military nuclear programmes also demonstrate some important facts about the rise and fall of the policy of ideologies or myths in the policy process. The distinction between civil and military nuclear programmes has been emphasised at least from the time when the first electricity-generating nuclear reactor at Calder Hall was opened. However, links between the two were not officially denied until the 1970s. Since that time the nuclear industry and government have sought, through secrecy, misleading statements, institutional changes and continual repetition, to sustain a myth which could be contradicted through information about the earlier nuclear programmes that is in the public domain. Myths and ideologies are sustained by social and political processes and can co-exist with contradictory evidence. It was only in the context of the rise of the anti-nuclear movement and increasing parliamentary concern about the economics of nuclear power that researchers began to piece together

Introduction

this evidence and, most importantly, to make effective political use of it. It is also interesting to note the rise of an alternative myth for radicals propounded by Tony Benn. As the chapter illustrates, rather implausible assertions about the British Government's motivations in switching to pressurised water reactors were sustained by the authority of an ex-energy minister and a section of an interest group that wanted to believe them.

David Lowry's chapter also raises questions about the effect of military links on the management of the British civil nuclear programme in the late 1950s. The decision to expand rapidly the Magnox reactor programme, made in the face of opposition from the Central Electricity Generating Board, was taken in the context of efforts to bolster Britain's great power status through military nuclear programmes. At almost exactly the same time Britain tested its first H-bomb and was also on the threshold of securing a nuclear exchange agreement with the United States. The expansion of the British nuclear programme and the prospect of large-scale generation of plutonium were probably thought to be an inducement to congressional acceptance of the Mutual Defense Agreement. Similarly, the Anglo-American Mutual Defense Agreement established an atmosphere and an institutional framework in which Britain was less likely to switch to light water and pressurised water reactor technology in the late 1950s and early 1960s than other European countries, such as France. In retrospect, the British decision to continue with indigenous gas-cooled reactor technologies was a very expensive one.

Embarrassment or secrecy about links between civil and military programmes may lead to other management problems. There is a tendency for military considerations to override civil ones where the two conflict, so great is the ideology of national security. Where no explicit conflict is perceived, military considerations, or some half-conscious idea of possible future military considerations, may distort decision making in the civil area. Distortion is probably greatest where the connection is a covert one. As the secret British Chevaline programme illustrated, political embarrassment, secrecy and a resulting government impulse not to get close enough to get its hands dirty, do not make for the sound management principles of cost control and accountability.

The interactions between civil and military electronics, the subject of chapter 5, are, though apparently less covert, just as difficult to disentangle. Rapid advances in electronic technology have become equally crucial to economic prosperity and high-technology weapons. Although there is likely to be a two-way interaction between advances in military and in civil electronics, disproportionate attention has been focused, both in research and in policy terms, on the idea of 'spin-off' from military to civil electronics. Clearly this is a strongly ideological

Introduction

area, since the proved existence of 'spin-off' in this direction is a way of legitimising existing levels of defence spending.

Although a number of studies have been made, particularly of the effect of military programmes on the British electronics industry, few of their findings have fed into official policy making, which in this, as in other policy areas, tends merely to select those research findings which support a particular policy or ideology. The idea of managing the interactions between civil and military electronics, in the interests of jointly optimising both areas, obvious as it would seem, is apparently far removed from reality on the ground. Departmental boundaries within the Cabinet and the civil service, with their distinct and prescribed budgets, tend to promote an adversarial rather than a collaborative relationship.

The British Ministry of Defence is, like defence departments elsewhere, charged specifically only with defence and security responsibilities, yet, as British industry's largest single customer, its decisions have a far-reaching effect on the domestic economy, the formal responsibility of the Department of Trade and Industry. The final chapter, on 'British defence decision making: the boundaries of influence', raises other questions about the management of complex military/civil interactions. The Ministry's procurement programmes operate within a policy of protection for the British defence industries. There is, however, as the Nimrod case illustrated, increasing conflict at the high-technology end of procurement between the preferences of the armed services on the one hand, favouring the cheaper, well-tested and compatible defence equipment produced by the Americans, and the pressure groups of British defence industries on the other. The traditionally secretive and élitist processes of the Ministry of Defence—keeping decision-making, research and information in house— have been connected, among other things, with a perceived need for it to shelter itself from the pressure of the domestic internal lobby.

However understandable, such closed processes of deliberation have associated, and possibly increasing, costs. The Ministry of Defence, conscious, after the shattering of the defence consensus, of the dangers of isolation, has indeed made some attempts to extend its contacts with the academic defence community. But the full range of available expertise and ideas has not been fed into British defence policy processes in the way its complexity might be thought to require, or even in the way it is said to do elsewhere. The domestic environment in which British defence decision makers operate in the 1980s demonstrates a greater interest in and knowledge of defence, as well as being more divided, than it has at any time since the Second World War. Decision makers, chapter 6 suggests, are still more likely to see this as a threat than as a resource.

4

Nuclear Weapons and Nuclear Power: Bias and Mythology in the Making of the British Magnox Nuclear Reactor Programme

DAVID LOWRY

Introduction

The management of defence policy in the pre-nuclear period was liable, even in times of peace, to intrude on other aspects of national decision making. The maintenance of British naval supremacy in the pre-First World War period, for instance, had implications for national policy towards merchant shipping, such as the Merchant Shipping Act of 1906. This confined pilots' licences to British subjects and prescribed better food and accommodation under the Red Ensign in order to retain British crews.[1] However, in the pre-nuclear period, the implications of defence policy for other aspects of national policy in peace time were both limited in scope and mostly overt. Since the advent of nuclear weapons, the interactions between defence policy, scientific policy, industrial policy, energy policy and parliamentary accountability have not only become more dense, they are also far more often unacknowledged, if not covert. The secrecy which shrouded the development of nuclear weapons from their inception has spilled over, in one form or another, into these other related areas. Successive British governments, Labour and Conservative, aware that the links between nuclear weapons and nuclear energy were potentially contentious, allowed the myth that they were quite separate to develop uncorrected during the 1960s and 1970s. The Government proved evasive and obfuscatory when nuclear links became a public issue of the 1980s.

The uncovering, in the 1980s, of evidence about the links between the development of nuclear weapons and nuclear energy in Britain in the 1950s, destroyed a myth which had been widely believed, even by well-informed and directly involved people. This evidence suggests that military considerations and the Anglo-American special nuclear relationship had an important influence on the development of British energy policy. This influence—rarely publicly acknowledged—reinforced an existing bias, stemming from exaggerated faith in big science and high technology, towards nuclear energy options at the expense of non-nuclear alternatives. It also distorted the calculations of the expected costs of nuclear-generated electricity. The military connection helps to account for British persistence in developing advanced gas cooled reactors for ten years after the French had switched to the more commercially advantageous American-type pressurised water reactors (PWR).

This chapter explores the influence of military considerations on the development of British energy policy in the period 1955–59, and the gradual exposure, in the period 1981–87, of the myths about the separation of nuclear weapons and nuclear energy which had been allowed to develop in the meantime. Even some of those who had formerly held high positions of responsibility claimed, in the 1980s, not to have known what had been going on. At the time of the public inquiry into the proposal to build a Pressurised Water Reactor at Sizewell on the Suffolk coast, Tony Benn, former Secretary of State for Energy in the Labour Government of 1974–79, claimed in his column for the *Guardian* in November 1983:

> It took me eight years, as the responsible minister to piece all this [a whole nexus of civil/military linkages] together but I am absolutely certain that ... it has been the prime purpose of the military establishment, over many years, to see to it if they could, that such an interpretation of what was happening should never be allowed to be understood by the public at large. ... This whole strategy has been covered up by the most massive Anglo-American public relations campaign about Atoms for Peace and Beating Swords into Ploughshares ...[2]

Benn's statement, which had a political origin and was published for a political purpose, contained the seeds of new mythologies. He claimed, almost certainly wrongly, that 'much of the plutonium which will be produced at those PWR stations will be separated out at the reprocessing plant at Windscale and then sent to the United States to be used to make the warheads for cruise missiles.' This chapter sets out to distinguish fact from fiction in an area where misconceptions, accidental or deliberate, have flourished. It looks at the implications of the mystification surrounding the plutonium link for standards of constitutional accountability and of policy formulation in Britain. It begins by exploring crucial periods in the history of nuclear developments in the United States and the United Kingdom, history which has only recently been pieced together. This provides a basis for disentangling myths, establishing linkages and exploring their implications. (A summary of the nuclear agreements between the United Kingdom and the United States is provided in Appendix 1.)

American nuclear development to 1955: civil–military links

Although American nuclear development during the Second World War had benefited from the inputs of scientists from a number of foreign countries, the United States Government hoped initially that there would be little nuclear development elsewhere. Most, though not all, of the Anglo-American wartime nuclear collaboration was abruptly terminated shortly afterwards.[3] The difficulties involved in trying to contain nuclear developments abroad were recognised in the United States as early as 1946, in the Acheson–Lilienthal report, and underlay the abortive American Baruch plan. By 1953 the American Government had bowed to the inevitability of nuclear development overseas, at least in civil areas. The Eisenhower 'Atoms for Peace' programme tried to manipulate American assistance with nuclear technology to overseas countries in ways advantageous to the United States. In most cases this meant providing assistance with nuclear energy programmes in return for promises of the strict confinement of developments to the civil sector, in ways intended to be also of commercial benefit to the United States. The case of Britain was an exception. In this instance the Eisenhower Government became eventually persuaded of the military advantages of nuclear collaboration. However, Congress was fearful that this would bring commercial advantages to Britain at the expense of the United States.

The existing nuclear powers hoped to prevent the 'proliferation' of nuclear weapons to new states hitherto without the prestige of a nuclear strike capability by maintaining a strict separation between civil and military applications. However, although American administrations publicly made much of the distinction between atoms for peace and atoms for war, nuclear energy was developed in the United States in close association with military programmes. In Britain, too, nuclear energy was developed under the aegis of the United Kingdom Atomic Energy Authority (UKAEA), which was responsible for the nuclear weapons programme after it took over atomic responsibilities from the Ministry of Supply in 1954. Britain, like the United States, was eventually to enjoin strictly civil development on other countries, while locked into her own internal nuclear connections.

The United States Atomic Energy Commission vigorously promoted the development of nuclear power, even though electricity costs in the United States were only about one-third of those in the United Kingdom and elsewhere in Europe, making its commercial justification questionable. Out of a wide choice of possible reactor designs, the Commission came to favour the development of the light water reactor, initially for military and then for civil development. The technical and political origins of this reactor technology in general, and pressure water technology in particular, have been much studied.[4] According to R. Perry:

> For practical purposes, the national [United States] commitment to the pressurised water reactor was made in August 1950, when the then Captain Hyman Rickover concluded that a proposal originated by the Oak Ridge National Laboratories and taken up by Westinghouse was the best prospect for the development of a power reactor for submarines.[5]

The development of pressurised water reactors was launched by Rickover in 1947 and was directed towards military uses in intercontinental bombers, surface naval vessels and submarines.[6] However, the United States was also capable, in this period, of following the alternative routes of developing power reactors using either natural or enriched uranium fuel. They were in an excellent position to play the game of 'design yourself a reactor'.[7] They began research on nearly twenty generically different types of reactor in all.[8] During 1951–52 the Atomic Energy Commission reassessed its approach to reactor research and development. For a time, with an eye on the escalating problems of the Korean War, it favoured dual purpose plutonium/power producing reactors. Ultimately, however, the Commission did not support the development of such reactors.[9]

In 1951 the Atomic Energy Commission began to encourage industrial participation in the development of atomic energy. In particular, it wanted a commercial assessment of the technical and economic feasibilities of the various atomic power plants on the drawing board.[10] Rapid strides towards commercial involvement—and indeed a change in the whole atmosphere surrounding atomic power in the United States and elsewhere—came with President Eisenhower's 'Atoms for Peace' announcement of December 1953, which implicitly acknowledged the impossibility of confining nuclear development to the United States. If the spread of atomic technology was inevitable, then it was better to give it benign connotations. Eisenhower contrasted the peaceful and the warlike potential of the atom and attempted to confine nuclear developments outside the United States to the purely civil. To this end, the United States offered 20 tonnes of enriched uranium to prospective beneficiaries overseas, for strictly civil purposes only,[11] as well as 20 tonnes of enriched uranium for the domestic development of civil atomic energy. Congress responded positively to the presidential initiative. After considerable horse-trading between Republican and Democratic members of the key Joint Congressional Committee on Atomic Energy, the Senate passed a considerably amended Atomic Energy Act on 16 August 1954. This permitted the private ownership of nuclear power plants in the United States for the first time and made possible executive 'agreements for co-operation' in the development of strictly civil nuclear developments with foreign states.[12]

Early in 1955 Rickover was in a position to commission the United States Navy's first pressurised water reactor propulsion unit in the submarine *Nautilus*, so named after Jules Verne's fantasy submarine. This reactor, with a power output said by Rickover to be equivalent to the electricity consumed by a city of 20,000 people,[13] was itself based on a prototype built at the National Reactor Testing Station in Idaho, which had gone critical in March 1953.[14] But, at about the same time, a wave of economy cuts by the Eisenhower administration led to the cancellation of another of Rickover's projects, a Westinghouse-designed pressurised water reactor propulsion unit for an aircraft carrier. Strong protests from both the Joint Commission on Atomic Energy and the Atomic Energy Commission persuaded the Appropriations Committees of Congress to restore funds so that Rickover could convert the reactor into a land-based power plant.[15] The British decision to build Calder Hall was a considerable stimulus to this decision: the Joint Committee on Atomic Energy was concerned that the British state industry would gain a decisive advantage in commercial nuclear power if the American project did not go ahead. The Atomic Energy Commission wanted an input of private capital to the development of the plant and utilities were encouraged to submit bids. The Duquesne Light Company was selected.[16] Construction began at Shippingport, Pennsylvania in September 1954, immediately after the passing of the Atomic Energy Act, and the reactor was commissioned in 1957, reaching full power in December of that year. The Joint Committee on Atomic Energy claimed full credit for its success.[17]

The pressurised water reactor had, in 1954, been considered to be the most conservative in design and least likely to produce competitive nuclear power of the five major lines of reactor development submitted in the Atomic Energy Commission's draft plan.[18] Four utility/industrial consortia responded to the Commission's call for proposals, and the resulting programmes, among them the failed Hallam sodium/graphite reactor,[19] met with varying success. The Atomic Energy Commission's second demonstration programme was also unsuccessful, a failure acknowledged by the Commission to have resulted from its treating immature reactor designs as if they were ready for near-commercial demonstration. (This mistake, born of a faith in high technology, was also subsequently made by the British ministers who accepted the Central Electricity Generating Board's advice on the prospects for the Dungeness B power station, incorporating the first advanced gas cooled reactor to be started.)

The American Atomic Energy Commission was so keen to get its demonstration programme underway that it invested lavishly in its support. By the end of the first six months of 1955, the total published capital investment in atomic factories was $6,600m. New construction in the fiscal year 1955 amounted to $870m.[20] It was in that year that decisions, based on military considerations, were taken which in effect helped to accelerate British civil reactor development and to push it along a different path from that of the United States, with major commercial consequences.

Anglo-American atomic relations, 1955–59

Although Anglo-American relations on atomic energy co-operation were less than cordial between 1945 and 1954, the years 1955 and 1956 marked a turning point in atomic energy development in the United States and the United Kingdom, and in Anglo-American relations. Although this change had implications for British civil developments, the stimulus for it was a purely military one. The success of the British atomic bomb tests, from October 1952 onwards, had an impact on the American administration's attitude to collaboration with the United Kingdom. Secretary of State John Foster Dulles saw the Atomic Energy Act of 1954, which permitted the sharing of 'external characteristics' of nuclear weapons and other military atomic information, as providing an opportunity to strengthen the nuclear component of NATO.[21] However, the desire of the American military to discuss various aspects of atomic collaboration with the British military ran into trouble with the Joint Committee on Atomic Energy, fearful that such collaboration might give their British competitors commercial advantages. The British, who had hitherto concentrated on developing the gas cooled (Magnox) reactor, might, Congress feared, use a military agreement to make possible their own development of the pressurised water reactor, in which the Americans had a decisive lead. As John Simpson records:

> The U.S. Navy's first nuclear submarine *Nautilus* had been handed over on 22 April 1955, its development having been pushed through by the then Captain Rickover, US Navy, with strong support from the Joint Committee. The Subcommittee on Agreements for Co-operation of the Joint Committee chose to regard nuclear submarine propulsion reactors as a new 'secret' of the atom, and during the hearings on the two Anglo-American agreements, members insisted that this technological lead should be preserved for purely national exploitation.[22]

Thus when the parallel Anglo-American civil and military exchange agreements were signed on 15 June, they included a provision in the civil agreement that prohibited any unauthorised discussion of pressurised water reactor technology.

The United States Strategic Air Command reached its maximum ever strength in Britain in the summer of 1955, when more than 11,000 American military personnel were deployed.[23] The Royal Navy decided to try to exploit the great co-operation which the British military establishment believed it was affording its American allies. It attempted to persuade the United States Navy to use its influence to get pressurised water reactor propulsion technology made available to the British. The first Sea Lord, Admiral Lord Mountbatten, visited the American Chief of Naval Operations, Admiral Burke, in November 1955 with this aim in view. He said that the uranium enrichment plant at Capenhurst would be available for submarine atomic fuel manufacture the next year, that the British wanted to secure an agreement, and that this would be to the mutual benefit of both parties. As Simpson explains:

> The visible result of his visit was that a British naval liaison officer joined the United States Polaris project team; the hidden consequences were that the United States Navy chose to drive through submarine technology exchanges with Britain, despite known opposition from the Joint Committee on Atomic Energy. Negotiations between the Department of Defense, the Atomic Energy Commission and Britain were therefore opened with the objective of drafting an amendment to the 1955 civil agreement, specifically to authorise exchanges of information on nuclear propulsion technology. The Joint Committee on Atomic Energy was kept in ignorance of this move, and to exacerbate the situation, the civil agreement, unlike the military one, specified that all exchanges were to be on the basis of reciprocity; yet Britain had no information to exchange, as it had undertaken little research and no development work on this type of reactor.[24]

The Joint Committee on Atomic Energy was predictably hostile to the plans of the United States Navy, knowing as they did that the United States was not going to gain any useful information on Britain's own civil reactor programme, the Magnox gas-cooled reactor. This was because the intellectual property rights for this programme were owned by the industrial consortium building the reactors for the Central Electricity Authority, not by the British Government. Yet the 1955 Atomic Energy Agreement stipulated that such exchanges must have an element of reciprocity: but if the United States was already developing a programme, this could mean that it had little to learn from elsewhere. Simpson describes this as a subtle 'catch-22'.[25]

Nevertheless, despite the Joint Committee's disquiet, the negotiations to transfer American pressurised water reactor submarine propulsion technology to Britain were concluded in June 1956.[26] Several members of the Joint Committee were outraged. Commissioner Murray attempted to halt the arrangement on 25 June and a report of the Committee on the security implications was published on 23 July with the same aim in view. However, the impending closure of the congressional session prevented implementation of the Joint Committee's recommendations. By the time Congress reconvened on 3 January 1957, a number of major political changes had occurred: on the international scene, the Suez war and the Soviet invasion of Hungary; in the United States, the presidential election; and in Britain, the opening of the Atomic Energy Authority's Calder Hall (Britain's first dual purpose military/civil plutonium production Magnox reactors near the Windscale site). President Eisenhower, respecting congressional desires, did not act on the draft submarine agreement during the congressional recess but, after his re-election, he informed the Joint Committee on Atomic Energy that the Secretaries of State and Defense and the Chairman of the Atomic Energy Commission had been instructed to initiate the exchanges with Britain.

British historians agree on the long-term importance of this outcome for the Anglo-American special relationship. Simpson comments that:

> The conflict over the 1956 amendment to the Anglo-American civil agreement was, in retrospect, a very significant turning point in the relationship between the Joint Committee on Atomic Energy, the Atomic Energy Commission, the Defense Department, the State Department and the administration. For the first time since 1946, the four latter organisations had banded together in a determined attempt to open up an exchange of nuclear information with Britain in the face of active opposition within Congress. The tactics used were rather crude, not to say opportunist, but they underlined a new-found determination to evade the congressional stranglehold on those international actions on nuclear energy which the administration judged essential for national security.[27]

Rickover visited Britain in May 1957 with an offer from the United States government to provide Britain with one pressurised water reactor for submarine propulsion, together with the necessary fuel. The agreement was completed on 27 May. The advantages to the British were clear, since they lagged behind in expertise on this type of reactor for submarine propulsion. The Admiralty had set up a naval section at the Atomic Energy Research Establishment, Harwell, only in 1954 and design development had begun only in 1956. Since highly enriched uranium was not expected to be available before 1960, there was no point in starting earlier.[28]

A group of 25 leading engineers from the United Kingdom Atomic Energy Authority, eager to remedy these deficiencies, visited the United States in June 1957 to learn at first hand from the Atomic Energy Commission and from Westinghouse about the pressurised water reactor's capabilities. The visit produced further acrimony, as members of the Joint Committee and other members of the American atomic establishment complained that the British delegation was using the guise of the submarine propulsion agreement to gain commercially valuable insights, especially into the Westinghouse operations. As Simpson notes, 'the link between the American submarine reactors and their land-based power reactor programme made such suspicions almost inevitable'.[29] Despite the fact that the Atomic Energy Commission Chairman, Admiral Lewis L. Strauss, had protested repeatedly that the United States had no need to 'run a kilowatt race with Great Britain', as the American research and development programme was in the long run much superior to the British effort,[30] these political concerns persisted and affected relations on the naval reactor negotiations.

It is not surprising that some of the Joint Committee members believed that Britain could be covertly seeking commercial gain under the guise of military negotiations for the submarine pressurised water reactor. The British White Paper of 1955 envisaged the expansion of nuclear energy and the British Electricity Authority found that there was no shortage of willing engineering recruits to its expanding atomic branch. Even earlier, in 1954, the United Kingdom Atomic Energy Authority, a powerful quango whose chairman had direct access to the Prime Minister and which enjoyed prestige because of the role it played in the development of the British nuclear bomb, was already beginning to press for the rapid development of nuclear energy. The UKAEA, anxious to extend its high technology empire, anticipated the conclusions of the Trend report on atomic energy, and had begun to encourage four leading turbo-alternator manufacturers each to get together with one of the four leading boiler manufacturers to form consortia which could handle the prospective new atomic power plant contracts.[31] The Central Electricity Authority, obliged only to pursue the most efficient generation of power for the grid, found themselves, throughout 1955 and 1956, increasingly pressured by the Atomic Energy Authority and the consortia to accept an increase in the nuclear programme above the already agreed levels of 1,700 megawatts (MW) by 1965.

Lord Citrine, for more than 10 years chairman successively of the British Electricity Authority and its successor the Central Electricity Authority, found himself increasingly seduced by the political advantages which Sir Edwin Plowden, the chairman of the United Kingdom Atomic Energy Authority, was projecting for increased atomic power development. In the early months of 1956, Plowden's assiduous lobbying of Whitehall resulted in a tacit agreement to expand the Trend projections for atomic energy and Citrine agreed to a near doubling of the programme to 3,200 MW. The Atomic Energy Authority exploited the Suez War for all it was worth, as a symbolic threat to future fuel security. The Government, encouraged by the announcement by the Central Electricity Authority in December 1956 of its first two nuclear contracts, decided in March 1957 to increase the programme to a target of 6,000 MW.

The Central Electricity Authority was taken aback at the magnitude of the increase envisaged. As Hannah comments:

> It seemed to the CEA to go beyond the bounds of common prudence, and they were appalled to find the new Minister [of Power] Mills seemed to have quite an inadequate appreciation of its far-reaching financial implications.[32]

Altogether, the British were seen to have made a substantial material and symbolic atomic commitment for the future. The lingering question for the doubting American congressmen was, could the British cope with just one thermal reactor system (the Magnox) under development? The Joint Committee were suspicious that the British were keen to pick up what they could about the pressurised water reactor. They saw the United Kingdom as their main commercial rival who would develop pressurised water reactors herself, should they prove more commercially viable than Magnox.

However, on the weapons side, Britain had proved its atomic virility[33] with the testing of its first H-bomb in the Grapple-1 Test on 15 May 1957, followed by two other H-bomb tests by 27 June.[34] The highly fraught and complex political negotiations between the United States executive and the Joint Committee that followed, coincided with diplomatic negotiations with the British to consolidate a new atomic rapprochement and are explained in detail by Simpson and Hesketh.[35] The purpose was to achieve what Prime Minister Harold Macmillan described to Parliament on 9 May 1959 as 'our interdependence in defence matters'.

The deep-seated worries within the American Congress about the effects of such interdependence upon the commercial property rights of American light water reactor research persisted. How was this issue resolved?

The whole debate over bilateral atomic relations with Britain was given added urgency by two events: the successful launch of Sputnik by the Soviet Union on 4 October 1957[36] and the fire that irreversibly damaged a Windscale plutonium pile on 8 October. Although in theory the piles could have been reactivated, especially the undamaged one, this was considered unnecessary in view of the imminent opening of the four dual-purpose plutonium and power production reactors at Chapelcross. Ten days later Macmillan, Plowden and the Permanent Secretary at the Ministry of Defence, Sir Richard Powell, held a meeting at which it was agreed that the British should rejoin the Americans.[37] The countries of the 'free world' should pool their resources to meet what they claimed to perceive as the increased threat from the Soviet Union. The following week a three-day meeting between Macmillan and the Foreign Secretary with Eisenhower came to an agreement. The joint communiqué ('A Declaration of Common Purpose') included the provision that:

> The President of the United States will request the Congress to amend the Atomic Energy Act as may be necessary and desirable to permit a close and fruitful collaboration of scientists and engineers of Great Britain, the United States and other friendly countries.[38]

It further noted that a study group headed by Strauss (Atomic Energy Commission) and Plowden (United Kingdom Atomic Energy Authority) was to make recommendations in the field of nuclear relationship and co-operation between the two countries.

Macmillan was in a delicate position at this time, for shortly after his return from Washington he was presented with the full technical report on the Windscale fire by Sir William Penny, the head of the Atomic Weapons programme for the United Kingdom Atomic Energy Authority. On 31 October, Macmillan decided to withhold part of the most sensitive technical details in Penny's report. In a Parliamentary statement a week later, on 6 November, he explained that the report was 'a technical document dealing with the design and operation of a defence installation [and] it would not be in the national interest to publish the report [in full]'.[39]

It was assumed that the secrecy was primarily due to Macmillan's concern not to divulge details of Britain's specialist military atomic research to the Soviet Union; and perhaps also to a desire not to give ammunition to the growing anti-nuclear weapons campaign. Both were probably significant factors. We know from cabinet committee meeting minutes released on 1 January 1988 under the 30 year rule that there was concern about potential exploitation of the accident by anti-nuclear weapons campaigners.[40]

But the most interesting revelation was that Macmillan was primarily concerned to keep details of the Windscale accident secret from the Americans. This is made clear in a minute from the report of a meeting of the United Kingdom Atomic Energy Authority on 4 November. Sir Edwin Plowden stated:

> He [the Prime Minister] thought even if it had been considered that there was no security objection to the publication of so much technical detail, there would still remain the danger that it would be quoted out of context and misused in other ways by hostile critics.
>
> In particular it would provide ammunition to those in the United States who would in any case oppose the necessary amendments to the McMahon Act which the US authorities proposed in order to make possible the desired degree of closer collaboration between the two countries in the military applications of atomic energy.[41]

Thus it seems that the Macmillan Government allowed it to be assumed that the reason for secrecy was security *vis à vis* the Soviet Union; but the major reason was in fact the desire to protect the beckoning special nuclear relationship with the United States.

The United States administration was then preparing to obtain the requisite congressional approval for the atomic arrangement with Britain. This was set in motion by Eisenhower's State of the Union message of 9 January 1958 in which he stressed that it was wasteful for a country to consume talent and money in solving problems which a friendly ally had already solved. Later, on 27 January, two bills incorporating the proposed amendments were introduced into both Houses of Congress. These were then taken up by the Joint Committee's Subcommittee on Agreement for Co-operation, which also planned hearings on the proposed co-operation with EURATOM, the nuclear agency of the newly founded European Economic Community.

Between January and May the Subcommittee held 12 full-day hearings,[42] which generated 530 pages of written testimony and appendices, plus one third as much again of classified testimony. These hearings have, as explained elsewhere by Hesketh and Simpson,[43] considerable significance both for the British Magnox reactor programme and for Anglo-American atomic defence relations.

The debate in the hearings centred on the likely future plutonium requirements of the United States, as projected variously by the Atomic Energy Commission and the Department of Defense, and on the possibility of making up the shortfall of plutonium, which some expected, from America's allies in general and the United Kingdom in particular. The concentration on Britain meant that the issue of the transfer of *Nautilus* submarine atomic-propulsion data for the British *Dreadnought* submarine constituted a substantial, and unremitting, theme in the hearings. Discussion of the availability of plutonium from the British produced a number of detailed enquiries and answers. The Magnox design produced per kilowatt the greatest abundance of plutonium of all the reactor designs available. (According to a later paper by the current Central Electricity Generating Board Chairman, Lord Marshall, the comparative figures for equivalent kilograms, i.e. the same fissile worth of plutonium, is advanced gas cooled: 173 kg; pressurised water reactor: 270 kg; CANDU: 493 kg; and Magnox: 617 kg.[44]) It was reported that Joint Committee members had been advised, on a visit to Britain, that plutonium would be produced in great quantities from about 1965, the date by which the British were also hoping to develop methods to incorporate it as a fuel.[45] It is clear from the record of these hearings, that it was expected that the British Government would soon have substantial quantities of plutonium, that the United States was interested in purchasing such plutonium for weapons development, and that this was the motivation, at least as explained to the Joint Committee, for the atomic collaboration between the two countries. As Strauss, Chairman of the Atomic Energy Commission, said of such collaboration in his testimony:

> It would assist their civilian power program primarily, but this is not primarily to assist those programs. This is primarily to supply plutonium to us for our unrestricted use, which is to say, at present, our military use.[46]

It was felt to be particularly appropriate that the countries constituting EURATOM (which did not yet contain the United Kingdom as a member) who were to be defended by American nuclear weapons deployed in NATO, should produce plutonium for the American nuclear weapons programme:

> Commissioner Vance: We are now proposing to furnish weapons for the defense of these countries we are talking about [i.e. the EURATOM countries] and those weapons are going to require plutonium. It seems to me this is thoroughly consistent . . . (classified discussion) . . . that if we are going to supply weapons to be located on their territory for their defense—it is thoroughly consistent that the plutonium which they produce and sell to us could be used for making the very weapons that they want.[47]

In the event, there were possible objections to plutonium from experimental research reactors made available to European countries and Japan under the 'Atoms for Peace' programme, being used for military purposes: the State Department objected that this would undermine the peaceful nature of the initiative.[48] No such objections could be raised to an arrangement with Britain. As Simpson points out:

> The diplomatic advantage of an arrangement [solely] with the United Kingdom was that its reactors were not part of either the 'Atoms for Peace' programme or the EURATOM agreement.[49]

The greatest political difficulty which arose in the Joint Committee hearings was in connection with attempts to change section 55 of the 1954 Atomic Energy Act, to enable the Atomic Energy Commission to purchase British plutonium on the basis of a long-term agreement. Eventually the difficulty was circumvented by the establishment of a $200m 'revolving fund' administered by the Atomic Energy Commission, which would enable the barter of fissile materials to be undertaken over a phased period complementary with the operating regimes of the reactors producing the plutonium.[50]

When Macmillan visited Washington on 7–8 June 1958—one week after the Joint Committee hearings were completed—he was conscious that, unless the amendments to the 1954 Atomic Energy Act were passed into law by Congress before the end of the first week of July, the Congressional timetable, with its long adjournment from August, would prevent the law from being enacted for another six months. It was consequently important that his meeting with the President should result in at least a provisional military agreement.[51] The Joint Committee's changes to the bill were passed by the full Congress in late June and the President signed the bill on 2 July. The Anglo-American atomic energy agreement for the mutual benefit of the defence of both countries—known as the Mutual Defense Agreement—was formally concluded in Washington the next day. Simpson summarises the importance of the event:

> Thus the ambitious [British] civil production programme directly assisted in the transformation of Britain's military nuclear position.[52]

In Britain, the Central Electricity Generating Board was far from happy with the expanded nuclear power programme which it had been persuaded to accept. At the same time as the delicate final negotiations were under way in Washington, Sir Christopher Hinton, its Chairman, unwilling to allow the CEGB to become too dependent on new generating technology, was battling in Whitehall for a reduction in the size of the Magnox reactor programme which the Board had been told to develop. He made it clear that he expected to play a major role in the future planning of the national supply system. As Hinton has subsequently explained:

> I was profoundly unhappy about the then enlarged atomic energy programme. I fought against it as hard as I could within the Atomic Energy Authority but I was no longer a 'deck officer'. Obviously, even though I was trying to be discreet, I was more at liberty to say what I thought about the nuclear power programme. At a lecture I gave to a conference [in June 1958] I criticised—if that is the right word, perhaps it is better to say commented on—what I considered to be an inflated programme. Mills [the Minister of Power] was furiously angry with me about this. I said, 'Well, it is what I think, you had better think carefully about what I said.'[53]

Hinton was eventually able to persuade the Government that the target date for installation of the first 6,000 MW of nuclear capacity should be postponed from 1965 until 1966. Publicly the Government announced this as a 'minor modification' and a deferment of the acceleration which was planned in 1957, but Hannah[54] suggests that it amounted to a 20 per cent cut in planned investment. However, although it was reported from a Federation of British Industry conference on nuclear energy that spring that 'a chill wind began to blow over the glamour and hoo-ha surrounding nuclear developments',[55] no Magnox station was cancelled.[56] In spite of the emerging commercial doubts, government policy was fixed[57] and all seven Magnox stations were given the go-ahead, before any working experience with reactors larger than Calder Hall had been gained.[58] Thus a massive public investment was made in a substantially untried technology, with major opportunity costs for other kinds of nuclear reactor, not to mention alternative forms of energy provision. The scale of public capital investment, prematurely legitimised, meant that nuclear power in Britain would be virtually inevitable until the end of the century (and, incidentally, large volumes of nuclear waste would also be inevitable, its effects lasting for many centuries thereafter).

However, although the drawbacks of the British Magnox programme were to become more evident as time passed, the military balance sheet looked good. Anglo-American atomic relations improved during the ten months following the 1958 Mutual Defense Agreement, as the international debate about nuclear weapons testing drew the British and the American military nuclear establishments closer together than ever in protective ranks. The Mutual Defense Agreement allowed Westinghouse to establish a commercial foothold within the British atomic energy community, as they were the contractors for the pressurised water reactor submarine unit, which was the key technology to be transferred under the agreement.[59] By late 1958, a British breakthrough in thermonuclear warhead design meant that it would not now be necessary to plan for the manufacture of additional supplies of plutonium for warheads in the Central Electricity Generating Board's reactors.[60] But in early 1959, with British military planners uncertain what to do now that technical choices had to be made about future warhead manufacture, the Atomic Energy Authority stressed the benefits of the fissile material barter arrangements discussed at the Joint Committee hearings the year before. Thus, in the spirit of Eisenhower's stated desire for the sharing of resources, it was formally agreed on 7 May to amend further the Mutual Defense Agreement,[61] adding a key section, number III (bis), which read:

> The Government of the United Kingdom shall transfer to the Government of the United States for military purposes such source, by-product and special nuclear material, and equipment of such types, in such quantities, at such times prior to December 31, 1969, and on such terms and conditions as may be agreed.

This amended paragraph should be read in conjunction with paragraph C of Article V of the original 1958 treaty, which states:

> Except as may be otherwise agreed for civil uses, the information communicated or exchanged, or the material or equipment transferred, by either Party pursuant to this Agreement shall be used by the recipient Party *exclusively* for the preparation or implementation of defense plans and in the mutual interests of the two countries. (Emphasis added.)

The word 'exclusively' has two important implications: it ensured that the United States used the nuclear material or equipment received from the British solely for military purposes, and hence could not directly provide commercial benefit; and it reinforced the agreement that the British should not themselves gain commercially in light water reactor technology as a result of the *Nautilus* deal.[62] The British were thus encouraged to stay with the development of what was to prove to be their commercially unsatisfactory Magnox system. Any consideration at this stage of switching development to the American-type pressurised water reactor might have looked like a breach of the terms of the Mutual Defense Agreement.

Some may also interpret this exclusivity of the Agreement as encompassing, above all, the exclusion of atomic co-operation with the French. In 1955, when the civil atomic co-operation agreement was being negotiated, a group of senior atomic scientists in Britain were keen to renew old acquaintance with former French colleagues with whom they had worked in Canada in wartime, a decade earlier. But even though the French made overtures towards shared research on gaseous diffusion technology for uranium enrichment, this was turned down because of the delicate politics of the revival in Anglo-American atomic relations. So both in gaseous diffusion and gas-graphite (Magnox) nuclear technology, the French were forced to go it alone, or at least without the assistance of the United Kingdom.

The British press highlighted different aspects of the amended Mutual Defense Agreement. The *Daily Telegraph* noted a cost benefit:

> The immediate object is to enable both countries to produce nuclear weapons more cheaply than hitherto by avoiding costly duplication of production capacity.[63]

The Times also stressed the benefits of the pooling of resources, but noted that the distinction between civil and military uses had been eroded:

> The most important technical fact behind the agreement is that plutonium of civil grade—such as will be produced in British civil nuclear power stations—can now be used in weapons [and] . . . from an international point of view it appears to follow that any country with civil nuclear power stations and a plutonium separation plant will be able to make some form of nuclear weapon without need to design or operate their specifically military production . . . the distinction between civil and military uses of atomic energy is thus rendered less clear than it was.[64]

The author, *The Times's* science correspondent, appears to have misunderstood the fact that, in the early operating stages of a Magnox nuclear reactor, the fuel will necessarily be discharged after a low burn-up as the core is brought to equilibrium. Thus the discharged fuel would consist of a uranium/plutonium mix containing plutonium of an isotopic quality of 85 per cent or more Pu-239, which is approaching the quality of weapons-grade material.* However, *The Times's* defence correspondent correctly predicted that the new amended defence agreement was virtually 'the last nail in the coffin for the McMahon Act', that is, the very restrictive American Atomic Energy Act, which became law in 1946 after the bill sponsored by Senator Brien McMahon had been passed by Congress.

On the same day that the Anglo-American Defense Agreement was signed, the United States and France also signed a special agreement by which the United States provided up to 440 kilograms of enriched uranium for the testing and operating of a prototype nuclear propulsion unit for a French submarine already under construction.[65] However, most significantly the Franco-American agreement, unlike the Anglo-

---

*Weapons grade material is defined in the United States as 93 per cent or more plutonium: the British plutonium could be blended with 97 per cent plutonium from Savannah River, one of the United States's own military nuclear material production facilities, to give 93 per cent weapons-grade material.

American one, was not reciprocal in nature.[66] This had an important bearing on the civil reactor choice in each country:[67] the Americans, who gave the French comparatively little in terms of military co-operation, did not fear that this was an agreement which could be exploited for commercial ends. Consequently, the French, unlike the British, were not effectively warned-off developing pressurised water reactors.

A week after the amendment to the Mutual Defense Agreement came into force, the Prime Minister informed Parliament[68] that modifications had already been installed in the existing nuclear stations to facilitate the new agreement and that:

> Any plutonium which we may exchange with the Americans will come either from the civil nuclear power stations or from the Atomic Energy Authority's own reactors at Calder Hall or Chapel Cross.

As a result of the Mutual Defense Agreement, the 1957 Electricity Act governing the operations of the Central Electricity Generating Board had to be amended to allow dual purpose operations of its reactors. In spite of its earlier, frank statement about the provision of plutonium to the United States from British civil power stations, the Government appears to have been unwilling that the full reasons for the amendment should be made public. The amendment bill to change Section 2(7) of the Electricity Act was proposed under the public guise that it was solely necessary to give the Generating Board permission to produce radio-isotopes for sale from its reactors.[69] This was the first sign of a reticence, not to say secrecy, which was to become habitual.

In the political environment of the late 1950s and early 1960s, the governments of the day were able to set the political agenda in such a commanding way that the previously acknowledged sale of plutonium to the United States never became a public, as opposed to a parliamentary, issue. Although parliamentary questions were asked in 1958 and 1959 and debates held in 1960 and 1962, little public concern was shown about the nuclear materials production process. The rise from 1957 onwards of the Campaign for Nuclear Disarmament as a potent political force was associated with its concentration on the emotional issue of atmospheric nuclear weapons testing. Nothing more was heard in Parliament of the Mutual Defense Agreement nor of its effects until March 1963 when Richard Crossman was told, in an answer to a question put to Denzil Freeth, Parliamentary Secretary to the Minister for Science, that:

> No plutonium has been supplied to any country other than the United States of America for defence purposes, nor have any contracts for such supply been made.
>
> Plutonium *is* being supplied for military purposes under the terms of the Amendment of 7th May 1959 to the Agreement for Co-operation in the Uses of Atomic Energy for Mutual Defence Purposes. I cannot disclose the details.[70]

Since this information was not made use of politically—the provision of plutonium to the United States not being a public issue—it did not register in the public consciousness. The barely noticed 1963 statement in the House was to do nothing, over the ensuing 15 years, to forestall a widely believed public myth, found convenient by successive governments. The rhetoric employed to bolster the International Nuclear Non-Proliferation regime encouraged the belief that nuclear weapons and nuclear energy were entirely separate matters in the United States and the United Kingdom.

British Governments, Conservative and Labour, but especially the latter, for whom nuclear questions were particularly sensitive, were happy to let the myth go uncorrected. Although the information was always in the public domain, no-one took the trouble to research it until John Simpson wrote a memorandum for the House of Commons Expenditure Committee in February 1979.[71] However, Lord Hinton has since confirmed that, during his time in office, plutonium from the Central Electricity Generating Board's two earliest Magnox reactors at Berkeley and Bradwell was indeed supplied to the United States Atomic Energy Commission for unrestricted use (including military use), although not to the Ministry of Defence for its military purposes.[72] Specific accusations have also been made about the military use of plutonium from the CEGB Magnox stations at Hinkley Point and Trawsfynydd.

Major policy issues arise from the above account, which was pieced together only in the mid 1980s. The first of these is the issue of open government. Over nearly a quarter of a century, during which plutonium links with the United States were not on the political agenda, successive governments could only be accused of allowing misconceptions to go uncorrected, of not striving to direct public attention to the evidence already buried in the public domain. In the 1980s, however, the political agenda on nuclear matters changed dramatically with the ending of the two-party consensus on defence, the reinvigoration of the Campaign for Nuclear Disarmament, and the public inquiry into the CEGB's proposed new pressurised water reactor at Sizewell. In the face of a concerted campaign of questioning into military/civil nuclear links, official spokesmen now frequently resorted to evasion, if not deliberate deception.

The making and unmaking of mythology

The fact that the immediate post-Second World War military atomic energy project in the United Kingdom had built fixed research and development facilities as well as industrial process facilities—such as at Windscale—meant that when the United Kingdom Atomic Energy

Authority was formed in 1954 and took over the governance and guidance of atomic energy from the Ministry of Supply, there was an inevitable dual military/civil use of these facilities. This was freely acknowledged in the 1950s without causing any controversy, apart from a few disgruntled comments by Members of Parliament and the occasional editorial in *Electrical Review* and *Nuclear Engineering*. But in the mid and late 1960s, 1970s and early 1980s, it suited both Conservative and Labour Governments to encourage the public impression to grow that the kind of strict distinction between military and civil nuclear matters built into the nuclear non-proliferation regime, as embodied in the 1970 Non-Proliferation Treaty, was also maintained in Britain; and to allow the plutonium link built into the Mutual Defense Agreement to fade from public memory. This mythology could develop, without explicit deception being necessary, since there was very little questioning on this issue in Parliament or elsewhere.

From the United States side, the terms of the Mutual Defense Agreement were always potentially embarrassing, since successive administrations, ever since Eisenhower's Atoms for Peace programme, had emphasised the importance of separating the peaceful and the military uses of the atom. Two decades later, the Carter administration actually attempted to make these claims a reality. It went as far as halting the commercial reprocessing of spent fuel in the United States and placing severe restrictions on the export of certain civil nuclear technology, because of their possible links with the proliferation of military programmes.

The strength of opposition in the United States, especially from the electricity utilities' management, to linking civil and military development was apparent in the reception given to the Reagan administration's plans to ease the postulated 'shortage' of weapons-useable plutonium for its projected new weapons programmes in the early 1980s. These would involve 'mining' plutonium from the backlog of spent light water reactor fuel stored by the utilities operating commercial nuclear plants. In spite of the obvious advantages to the utilities of such an arrangement, which would transfer the responsibility for spent fuel and nuclear waste management from their shoulders to those of the federal government, both the heads of the utilities themselves and Dr Sigvard Eklund, Director-General of the International Atomic Energy Agency, thought it essential to retain the distinction between civil and military nuclear programmes, lest the whole panoply of carefully constructed international arrangements for safeguards against nuclear proliferation be terminally undermined.

The *New York Times*, alarmed in September 1981 by the disclosure that the powerful techniques of laser isotope-separation then under development to enrich uranium more cheaply, would also be able to separate bomb-grade plutonium from spent civil nuclear fuel, called on President Reagan to tell the world that the United States would not under any circumstance divert civilian nuclear power materials to its military programmes. The paper's editorial illuminated the power of existing mythologies:

> For a generation the peaceful atom has struggled, like a reformed prisoner, to live cleanly, on the sunny side of the street, shunning any association with its dark, brooding cousin, the bomb. This effort to go straight has been largely successful, in part because of American efforts. So it is jolting to hear the Secretary of Energy propose something that sounds like parole violation.
>
> The proposal is not just misguided but dangerous. It would facilitate making weapons; it also risks setting off a joint campaign, by people opposed to nuclear power as well as those opposed to nuclear weapons, in the United States and Europe.[73]

The United Kingdom Government, for its part, could increasingly expect public opposition to the supply of its plutonium to the American weapons programme, were this widely known, as bipartisan support for both nuclear power and nuclear weapons came under pressure towards the end of the 1970s. A number of apparently unrelated events combined to bring such issues on to the public agenda. In December 1979 the Amendment and Renewal of the 1958 Mutual Defense Agreement was completed in the United Kingdom. That same December NATO concluded its 'dual track' agreement on intermediate nuclear forces, to enable further American weapons to be based in the United Kingdom and elsewhere in Europe. These arrangements were followed shortly afterwards by the Trident Sales Agreement in January 1980—with terms specified to be as in the Polaris Sales Agreement—which also helped direct attention towards the issue of Anglo-American nuclear links. There was, moreover, the proposal of the Central Electricity Generating Board to build its first pressurised water nuclear reactor, based on an American design, at Sizewell.

Consequently, both the Government and the Central Electricity Generating Board, in the context of the Sizewell public inquiry, found themselves facing the kind of sustained and intensive questioning which had not been seen for a quarter of a century. The Sizewell inquiry, beginning in June 1982 with preliminary hearings at which objecting groups ensured that the issue of plutonium use and production would be aired, brought to a sudden end more than a quarter of a century of quiescence. The inquiry was to last for more than two years, and to produce 16 million recorded words and more than 5,500 submitted documents weighing in total 55 tons.

In January 1983, John Baker, the leading Central Electricity Generating Board policy witness, gave unequivocal evidence before the inquiry that:

Plutonium produced by the Central Electricity Generating Board reactors has never been applied to weapons use in the UK or elsewhere.[74]

Lord Hinton, the former CEGB head, interviewed one week after this statement had been accurately reported in the press, described it as 'deplorable' and said he considered it important that the Generating Board 'should not tell bloody lies in their evidence'.[75]

The Government too, as we have seen, was coming under pressure during the run up to the opening of the Sizewell inquiry, on the issue of plutonium links with the United States. In October 1981 the *Guardian* had reported that the United States had asked to buy British plutonium to meet the needs of its expanding weapons programme.[76] The Foreign and Commonwealth Office issued an official statement that plutonium exported to the United States would not be used for weapons but for the American fast breeder reactor programme. Although they did not challenge the truth of this statement, two concerned scientists challenged its import. Professor Sir Martin Ryle wrote to *The Times* (17 October) that:

> The plutonium, presumably from the spent fuel of Magnox reactors, is said to be destined for fuelling fast breeder reactors. FBRs produce extremely high-grade weapons plutonium and very expensive electricity. The plutonium they produce will comprise different *atoms* from those dispatched from the United Kingdom, but will depend on the latter for its existence.
>
> But even if the FBRs are only used to make electricity, the United Kingdom plutonium will enable United States plutonium production to be diverted to President Reagan's large new weapons programme.
>
> However innocently you try to present the transaction, the fact remains that United States weapons will be built which could not have been built without the aid of material produced in CEGB reactors.[77]

A letter to *The Times*, written on the laboratories' headed notepaper by Dr Ross Hesketh, a nuclear materials scientist then employed at the Generating Board's Berkeley nuclear research laboratories, indicates how prevalent was the myth of separate civil and military development; even concerned employees were unaware of the history of their industry:

> As a member of the civil nuclear energy programme of the United Kingdom I have for several years assured my critics that civil nuclear energy is distinct from military nuclear energy; I have assured them that Berkeley, Bradwell, Sizewell, Wylfa, have no connexion with the escalation of nuclear weapons ... If at this juncture the United Kingdom were to sell plutonium to the Reagan administration, I do not think that we, the United Kingdom, have distinguished civil use from military use.[78]

Hesketh was sacked in June 1983, then reinstated and finally forced into early retirement after a long struggle with his employers over his outspoken views on this issue.[79]

As David Fishlock, the knowledgeable and pro-nuclear science editor of the *Financial Times*, reported at the time, the Central Electricity Generating Board—preparing itself for the Sizewell inquiry—was decidedly nervous about British public reaction to what was said to be an American scheme to buy British plutonium.[80] Shortly after Fishlock's article the *Guardian* carried a full page analysis of the historic background and likely future complications of the Mutual Defense Agreement and the plutonium puzzle. This feature by Dr Norman Dombey, a physicist at Sussex University, was an important milestone in setting the agenda for research and debate thereafter.[81]

In the Spring of 1982, the Electrical Power Engineers' Association, the professional association to which most scientists working in the civil nuclear industry belong, publicly objected in principle to the use of plutonium from civil origins for military purposes. The Association, like Hesketh, appeared to be unaware of the history of its own industry; it threatened to withdraw its support for civil nuclear power if such diversion of plutonium to military ends was proved. This raised the stakes involved.[82]

However, the historical study by the present author and scientific analysis by four members of the British group Scientists Against Nuclear Arms in 1985[83] went some way towards clarifying matters. The latter's paper, published in *Nature* in September 1985, argued that 6.3 ± 0.8 tonnes of the plutonium which had been produced in British civil Magnox reactors, up to March 1985, had not been recorded in the official total for the inventory of the national civil plutonium stockpile. The authors commented that:

> It is interesting to note that our estimate for the balance agrees with the figure of 6.667 tonnes which was expected to be the maximum involved in the (1958/9) exchange between the United Kingdom and the United States, based on costs in the US enabling act.[84]

This paper had originally been prepared as part of research done for the Campaign for Nuclear Disarmament's Sizewell working group, which included the present author, Dr Keith Barnham of Imperial College, London, Dr Hesketh, and Rob Edwards, the co-author with Sheila Durie of *Fuelling the Nuclear Arms Race: the Links between Nuclear Power and Nuclear Weapons*.[85] This research team gave evidence throughout the public inquiry (June 1982 to December 1984), and was the final opposition organisation to present its closing statement. Apart from presenting its own evidence, the group also cross-examined witnesses from the Department of Energy, the Central Electricity Generating Board and British Nuclear Fuels Ltd. Much clarification of this complex issue was advanced through this procedure.

On 20 March 1986, a year after the Sizewell inquiry had closed, Lord Marshall, Chairman of the Central Electricity Generating Board, brought many years of misconception and a period of official denials to an end by saying in a television programme, that plutonium from the Generating Board had in the past been transferred to the military stockpile.[86] He denied, in a letter to *The Times* of 6 June 1986, that this was saying anything new.

There are many reasons for concern about the mystification surrounding government policies—a phenomenon particularly pronounced in the case of nuclear policies. One such reason is particularly important here. Where there is mystification or secrecy, the normal mechanisms designed to ensure accountability and the efficient management of publicly stated goals cannot operate. The vast cost overrun of the British Chevaline programme to provide a new 'front end' for its ageing Polaris missiles was connected with the continuing secrecy of its development under three different governments. How far can the manifold problems and over-optimistic cost predictions of the British nuclear energy programme be attributed to distortions resulting from its little known connections with military programmes? Did the military link produce a bias towards investment in nuclear energy rather than other, alternative sources of energy, and towards a less commercially rather than more commercially advantageous reactor system?

The energy policy issue

Recounting his knowledge of the fissile material barter arrangement in an interview with the present author, Lord Hinton raised the question of the plutonium credit, an issue which the former Atomic Energy Authority Chairman, Lord Sherfield (formerly Sir Roger Makins), had described in an interview with the author in October 1982 as unimportant. Hinton considered that the plutonium credit has implications for the management of the nuclear energy programme:

> While the initial industrial reactors were being built, the Atomic Energy Authority said that they would like them to be so designed that military-grade plutonium could be produced in them. The design was modified in such a way as to make this possible.
> The irradiated fuel elements were handed over to the Atomic Energy Authority ... [but] Plowden did not sell them to the Americans, he exchanged them for enriched uranium. I used to argue with him about this ... Plowden and I used to have dinner together once a month ... and I argued with him when he used to boast to me that he was getting very cheap enriched uranium from the States. I used to say, 'Maybe this is what it looks like, but in fact all it proves you are doing is not giving us a big enough credit for plutonium. By doing this you are disturbing the future of nuclear power, because future nuclear reactors will be using slightly enriched uranium and you will base your estimates of cost on the enriched uranium that you got cheaply from the States, simply as a result of not giving me a big enough plutonium credit.'[87]

Asked whether he thought that the original fuel cycle costs were distorted by this system of 'unfair exchange', Hinton replied:

> I do not know what quantities were involved so I cannot say to what extent there was distortion; but there was room for distortion.[88]

In other words, not only did the Atomic Energy Authority obtain, by two parallel but separate deals, highly enriched uranium for submarine use and highly enriched uranium for submarine and/or bomb use, but the economic appraisals of the advanced gas cooled reactor systems, then under prototype development by the civil research section of the Atomic Energy Authority's industrial group, may have been distorted in their favour as a result of the Mutual Defense Agreement. The civil nuclear power sector thus had two unplanned benefits as a result of the military atomic agreement with the United States. Apart from the latent consequence of the fuel-cost gain to the advanced gas cooled reactor — at least in appraisal terms — the CEGB's Magnox programme was also given a secret endorsement, as it was to partly provide for quantities of plutonium agreed for barter with the United States.

An example of this endorsement is the case of the Wylfa Magnox reactor. From 1961 onwards, the future of the final and largest of the Magnox reactors, at Wylfa in Anglesey, was the subject of considerable political debate. There was unease about the unhappy relations between the Central Electricity Generating Board and the Atomic Energy Authority and the effect which these had on the putative nuclear construction group. Professor Roger Williams in his 1980 study *The Nuclear Power Decisions* had described the 'sharp politics' between the Cabinet Office, the Treasury, the Board of Trade, the Central Electricity Generating Board, and the Atomic Energy Authority, following the Powell Committee* meetings and the parallel Parliamentary Nationalised Industries Committee inquiry of the same year (1962). The next year, the bitterness became public, not least in the Wylfa debate in the House of Lords in July, when Hinton was accused by Lord Coleraine — a strong supporter of nuclear power who considered that the British had made a series of misguided industrial decisions on nuclear power development — of 'economic dictatorship on a pretty wide scale'.[89] Hinton's interview adds a new, if incomplete gloss to the Wylfa debate, with a comment which can be interpreted in the context of the military Defense Agreement requirements. Asked whether in fact the CEGB might have been forced by the Government of the day to break the 1957 Electricity Act by taking on nuclear generating capacity when they admitted that the cost per kilowatt was

---

* The Powell Committee was an official review body under the chairmanship of Sir Richard Powell, then Permanent Secretary of the Board of Trade, investigating reactor choice options.

more expensive with nuclear energy, he replied:

> I think there is without doubt substance in this; and if you look at the evidence to the parliamentary Select Committee on Nationalised Industries in 1962, you will find a quantified statement by me of what nuclear power had cost the Generating Board.[90]

Asked whether, in his view, the Government was quite happy to allow him to continue to run the CEGB with a component of its generating capacity that was not economic, Hinton retorted:

> In fact, not merely was the Government happy but they forced us to fulfill the programme that they laid down. Wylfa is a long and sad story. It ought not to have been built at all, but when I suggested this to the Permanent Secretary at the Department, he said you have got to build it in order to meet the Government programme. There is no documentary evidence for this, but you can take it from me that it is true.[91]

Although no clarificatory question was put to Lord Hinton on the matter in this interview, the context indicated that the word 'programme' referred, not to the nuclear electricity generating programme as operated by the CEGB, but to the plutonium production programme agreed in secret with the United States.[92]

Policy implications

Three historical questions have implications for present day nuclear and high technology investment programmes. These are, first, did the use of the Magnox design for Britain's first 'civil' nuclear power programme make the disengagement of civil and military activities, as envisaged in Eisenhower's 1953 'Atoms for Peace' speech to the United Nations, a logistical impossibility? Second, how far did the early realignment with the United States on military nuclear matters, in the mid to later 1950s, set the diplomatic, defence and technological agenda for the British civil nuclear programme as well as for its military programme? Third, why was it that the French, whose scientific involvement, along with the British at Chalk River in Canada, provided a cadre of scientists with gas-graphite (Magnox) experience, decided to adopt the pressurised water technology in 1969, a full decade before any British minister made a similar endorsement? In other words, why did the British go along the Magnox route, taking 10 years longer than the French to realise the commercial advantages of the pressurised water reactor?

The French never had the same special historic and cultural relationship that the United Kingdom enjoyed with the United States. France has long taken relatively independent decisions on its strategic future.

The national humiliation of invasion and occupation in the Second World War reinforced the desires of France to establish itself firmly as a great nation again, a programme symbolised by General de Gaulle and Gaullism, characterised by its independence and centralising tendencies.

It is in this context that the role played by France in the establishment of the European Economic Community out of the former European Coal and Steel Community in 1957 needs to be understood. The present author has discussed this in detail elsewhere.[93] However, it is worth restating here that it was the French involvement in EURATOM that ultimately made is possible for American-type light water reactor technology to displace France's own gas-graphite technology. This was despite the great political power and prestige vested in the *Commissariat à l'Energie Atomique* (*CEA*), which, like the United Kingdom Atomic Energy Authority, fervently supported its own gas-graphite technology against the intrusion of the light water lobbyists from the United States.

The joint United States–EURATOM reactor programme, a loss leader for the United States permitting them to establish a foothold in Europe, gave the Federal Republic of Germany in the late 1950s and early 1960s the opportunity to establish a nuclear reactor programme under American sponsorship. This reflected the same ethos promoted a decade earlier in the Marshall plan for Western European reconstruction.

Once the French had agreed to participate in a joint nuclear reactor venture with the Belgians, for a light water reactor at Chooz on France's border with Belgium and Luxembourg, the French state electricity utility *EDF* (*Electricité de France*) was able to begin to build in-house expertise in this technology and was no longer totally reliant on guidance from the *CEA*. This ultimately created a bandwagon effect for the light water reactor in France. However, these commercial and technological developments have to be understood in the wider context of Anglo-French relations as they developed in the shadow of differing political relationships between the United States and the United Kingdom, and the United States and EURATOM.

Following the signing of the United States–EURATOM agreement mentioned earlier, the British Government was asked in a number of parliamentary questions about its implications for the British nuclear programme. In June 1959 the Prime Minister was asked to clarify the arrangements in the British agreement made with EURATOM for limited research co-operation. He replied that close contacts between the UKAEA and government departments were maintained with the relevant industries on the matter. In practice, Britain did not want any serious commitments within EURATOM; it preferred to opt for a separate, special nuclear relationship with the United States. However by March 1962, following vigorous negotiations undertaken by Edward Heath, Macmillan's government was ready to begin negotiations to join the EEC and EURATOM. De Gaulle, the French President, had to be won over to the idea. Thus Macmillan met de Gaulle in June. However, Macmillan was forced to make clear that:

> While we might co-operate in some of the details which were within our control, we could not part with these secrets which we only received from America as heirs of the original founders of nuclear science in the war.[94]

As the French had worked intimately with the British scientists at Chalk River during the Manhattan project, they shared the knowledge acquired there in any case. The real blockage on the British sharing atomic secrets with EURATOM, and hence France, was the exclusivity clauses built into the 1958 Mutual Defense Agreement and its 1959 amendments. Indeed, Macmillan tried to persuade the United States Administration that the special atomic secrets should be shared with France, if the French would sign an atmospheric test ban agreement.[95] But neither the French nor the Americans were interested in such an accommodation.

This unintended consequence of the Mutual Defense Agreement, and the Anglo-American special relationship, clouded relations between Britain and France, and Britain and the United States, for a number of years. On the one hand, it contributed to Britain's failure to join the EEC. On the other, it strengthened indigenous nuclear power technology (in effect, the prestige of the United Kingdom Atomic Energy Authority) against the predatory attempts of the American light water reactor salesmen, who could not use EURATOM as a means of entering the British market. In other words, Britain was the more inclined to stay with her own Magnox technology because she was not in EURATOM, and she was not in EURATOM partly because of the Mutual Defense Agreement with the United States. The consolidation of Anglo-American atomic relations with the 1959 amendment to the Mutual Defense Agreement was, in effect, the key 'moment of transition' that set the organisational and political power infrastructure which was to underpin civil nuclear power decisions for the next 20 years in Britain.

Thus it was that the French technocratic establishment was gradually weaned off the development of gas-graphite technology for the commercial expansion of nuclear energy—although the gas-graphite reactors were retained to provide plutonium for the expanded *Force de Frappe* atomic weapons programme. By the end of the 1960s, France was committed to the American light water reactor technology. This eventually opened up a worldwide nuclear reactor market—uninterested in the British Magnox system—to French penetration.

Just a few months after the great symbolic triumph of American technology, the moon landing in July 1969, the executive director of *EDF*, Marcel Boiteux, explained at the opening ceremony of the St. Laurent des Eaux gas-graphite plant, on 16 October, why the change to light water reactors was to take place.

> We have to acknowledge that a light water model is not more reliable than a gas-graphite model. . . . But the world currently has around 80,000 MW under construction or on order from light-water models, while there are 8,000 in service or on order from graphite-gas models. You see the disproportion.
>
> For France, within our little borders, to continue pursuing a technology in which the world has no interest doesn't make sense today. The fact that the world market is now clearly oriented towards light-water models means that our industrialists will only be able to enter the industrial world insofar as they have their own valid experience with the models the world is interested in.[96]

The British decision to stay exclusively with advanced gas cooled reactor development meant that she excluded herself from global nuclear development. The commercial logic which had convinced the French was also advanced in the reactor choice debate in the United Kingdom, but it took a further decade of trials and tribulations before it succeeded.

What is important here, above all, is to recognise that military exigencies had effects penetrating deeply into national energy policy. It was covert, rather than overt, civil military links that forged the power of institutions in the nuclear field; but in the public arena the two were allowed by government and the military to be perceived as separate. The military linkage, enshrined in the infrastructure, gave great prestige to civil nuclear developments and encouraged over-optimistic, if not deliberately falsified, costings for nuclear energy. This led to massive public investment in nuclear at the expense of other forms of energy. The first generation of gas-graphite reactors was developed, not because they were thought to be the most cost-effective generators of electricity, but because they, pre-eminently among existing reactor types, could create plutonium of the high purity necessary for military purposes. This factor, along with other military and geostrategic considerations, led to the prolonged development of the Magnox rather than the more commercially viable pressurised water reactor system.

The constitutional issue

How far were ministers, constitutionally collectively responsible for government policy during their terms of office, aware of the civil/military plutonium links between Britain and the United States outlined above? This chapter began with a quotation from the Rt. Hon. Tony Benn. While the present author disagrees with Benn's contention that much of the plutonium produced in the Sizewell B nuclear plant is earmarked for the United States's weapons programme, his other contention, that civil plutonium from other CEGB reactors has been exported to the United States for military use, is sustainable, as we have seen.

Let us briefly examine some of Benn's statements in Parliament concerning the Anglo-American deal on fissile material exchange and the extent of his ministerial knowledge of the details of such plans. On 12 January 1976 the left wing Member of Parliament and CND supporter Frank Allaun asked Benn, who was then Secretary of State for Energy, 'to which countries plutonium was exported from Great Britain, and was he satisfied that such exports were essential to our nuclear industry'. Benn's reply, both honest and evasive, reads as follows:

> Plutonium exports are not separately designated in the overseas trade statistics of the United Kingdom, but are subject to licensing control and international safeguards. I am satisfied with these arrangements and with the purposes for which these exports are undertaken.[97]

However, by December 1983, a month after his *Guardian* article, Benn expressed great dissatisfaction with the plutonium exports to the United States. As his testimony makes clear,[98] Benn now believed that plutonium from civil nuclear plants had been exported to the United States for military use, having been reprocessed at Sellafield.

Following the CND evidence to the Sizewell inquiry in November 1984, Benn began to put some questions on the American deal to the then Energy Secretary. On 3 December 1984 he asked if the British Government supported congressional legislation to restrict to civil use alone any plutonium exported to the United States from the United Kingdom under the Mutual Defense Agreement. The reply received read:

> We have had repeated assurances from the United States Government that it is not their policy to use plutonium exported from the United Kingdom under barter arrangements before 1971 for weapons purposes. The proposed legislation referred to by the Rt. Hon. Member is not a matter for Her Majesty's Government.[99]

In fact the United States administration had not said that it was 'not their policy', but rather had stated that they 'did not rely upon' the plutonium imported from the United Kingdom. Moreover, the Minister's reply did not address the possibility that plutonium of civil origin had been exported from the United Kingdom after 1971. These questions were examined in great detail under the cross-examination of CND's evidence at the Sizewell inquiry.

Benn did not drop his enquiries there. On 15 January 1985 he asked the Secretary of State for Energy in a further question whether he would publish any exchange of letters between Her Majesty's Government and the United States Government that constituted a waiver on the stipulation in the Mutual Defense Agreement that nuclear materials bartered under the arrangement should be used for the furtherance of defence purposes by each state. (The Mutual Defense Agreement stipulated 'exclusive' use in defence plans 'except as otherwise agreed'.)

The Minister's reply stated that it was not Her Majesty's Government's practice to release details of exchanges such as those between the United States and United Kingdom on the Mutual Defense Agreement.[100] Setting aside that this is not an entirely truthful answer, as the same Government did disclose some details of nuclear warhead tests in Nevada also conducted under the Mutual Defense Agreement provision, had any such 'waiver' in the form of an exchange of letters been published, it would have assuaged suspicion that the British plutonium had been put to military use. As the Government chose to decline to answer, this further fuelled suspicion that the denials of military use were only re-writes of history forced by the questioning and probing in Parliament.

It certainly would seem that Benn remained more convinced than ever about the truth, as he saw it, of the end use of the exported British civil plutonium. In his contribution to the debate on nuclear power in Parliament held on 13 May 1986—two weeks after the Chernobyl accident when there was much criticism of Soviet secrecy and cover-up—Benn repeated the charge:

> The biggest cover-up of all, for which I shall never forgive those responsible, was that throughout the period when I was Minister plutonium from our atoms-for-peace reactors was going to America to make bombs and warheads that would return to American bases here. That view has been confirmed by Ministers in this Government. I was cross-examined about it at the Sizewell inquiry, and only recently has it been admitted that the atoms-for-peace power stations are in reality bomb factories for the United States.[101]

Finally, in the parliamentary debate on the publication of the report by Sir Frank Layfield on the Sizewell B inquiry, Benn reiterated his personal doubts. These statements, made in Parliament on 23 February 1987, clearly constitute Benn's formal disowning of his earlier statement of the ministerial reply he made on 12 January 1976, to which reference is made above.

> I raise another point because of the bitter resentment that I feel. Throughout the whole period when I was the Minister no disclosure was ever made to me that the plutonium from these atoms for peace stations was going to America for weapon warheads. That information was denied not only to me as the responsible Minister who had a duty to convey to Parliament and the Public what was happening but to the House after I left office. When I was interrogated by Lord Silsoe on behalf of the Generating Board during the Layfield inquiry, I was rigidly cross-examined and asked three times to withdraw what I then knew to be true, because Dr

Hesketh, who was fired from the Generating Board for disclosing this information, said that it had happened.

I quote from one passage stating what happened on day 333 of the inquiry when the counsel for the generating board replied to Campaign for Nuclear Disarmament evidence. It states:

> the basis of the Board's position was to rely on Government statements and assurances that none of its civil plutonium has been diverted to weapons purposes.

That was a lie, yet it was said in the House and told to the public. It is on that basis that people still sometimes speak of 'peaceful purposes'. I am not trying to identify the Minister responsible. I do not suppose that he was told any more than I was, yet, according to the United States Secretary of State of Energy, the United States is required to use for military purposes all the CEGB plutonium received under the defence agreement.

If democracy is to survive, people must be told the truth.[102]

This brings us to the final question to be answered. Compare these two statements made by energy ministers in Mrs Thatcher's administration. The first is by Nigel Lawson who, when Secretary of State for Energy, told the Commons press gallery on 19 January 1983—a few days after the CEGB evidence on the same subject to the Sizewell inquiry that:

> There is confusion between nuclear power stations and nuclear weapons. I'm not going into the question of nuclear weapons now but there is no more connection between the generation of power in a nuclear power station and nuclear weapons than there is between a conventional power station and conventional weapons.[103]

Clearly this is nonsense, otherwise why the panoply of international safeguards and non-proliferation arrangements and agreements to deter the diversion of nuclear materials from civil power reactors to nuclear explosive purposes?[104]

Indeed, nearly four years later, the junior energy minister Alastair Goodlad admitted as much in a parliamentary reply to Dafydd Elis Thomas on 28 January 1987. He said in reply to a question asking why certain documents on the foundations of the civil nuclear programme in 1955 had been withheld from disclosure under the '30 year rule':

> It is well known that the civil and military nuclear programmes had common origins. The records in question contain sensitive information concerned with national security, the disclosure of which would not be in the public interest.[105]

There are two other legacies of secrecy arising from the dual military/civil role of the United Kingdom atomic energy programme. One is the long saga of the refusal of the United Kingdom government to permit access by EURATOM safeguards inspectors to the Sellafield reprocessing complex, which handles civil and military spent Magnox fuel together. This ultimately resulted in the European Parliament—which has a *locus standi* concerning oversight of certain arrangements and practices agreed between EEC member states and the European Commission on atomic energy matters—initiating complex procedures, after nearly three years of investigation, for taking the Commission and United Kingdom Government to the European Court. Whilst this issue also arose from information gained by CND in its involvement at the Sizewell inquiry in November 1984, the detailed developments may be found described elsewhere.[106]

The second legacy is that of the distortion inherent in the intertwining—on institutional and political levels—of one technology (nuclear weapons production), clearly covered by certain levels of national security, and a second (nuclear energy production), that need not be so restricted, but in practice has been. Elsewhere this author has detailed the full range of restrictions on information disclosure promulgated in the year following the Chernobyl accident, which was promoted as being one of new openness.[107]

Here we focus upon one aspect: the way in which the Government has changed the rules of parliamentary reply—thus undermining a crucial convention that makes ministers responsible to Parliament—when having to answer truthfully would show that earlier answers were untruthful. In brief the issue is as follows. On 4 February 1983—just after the CEGB evidence to the Sizewell inquiry on plutonium and the statement by Lawson cited earlier—John Moore, the then junior energy minister responsible for nuclear energy matters, told Parliament:

> No plutonium produced in any of the CEGB's nuclear power stations has ever been used for military purposes in this country and there are no plans to use it thus in the future. Further no plutonium from the CEGB nuclear programme has ever been exported for use in weapons.[108]

After persistent parliamentary questioning by a small group of MPs from the Labour, the Liberal, and the Plaid Cymru party over the next three years, Mrs Thatcher made the following statement to Parliament on 15 April 1986, one which is significantly different in its restricted scope:

> No plutonium produced in civil reactors in this country has been transferred to defence use or exported for such use *during the period of this administration*.[109] (Emphasis added.)

When asked specifically three days later if plutonium from civil reactors had been transferred to defence use or exported for such use during previous administrations, the Prime Minister replied that she was

> not able to answer for previous administrations.[110]

On 8 May 1986 the Prime Minister reasserted to Parliament that it is

> the convention that Ministers are answerable in Parliament only for matters

falling within the period of responsibility of the Administration of which they are members.[111]

This reply was in response to a question asking if the policy on answering questions on plutonium had changed since the all-encompassing answer by Moore cited above. The issue was further probed on 14 July 1986 when the Prime Minister was asked if this convention precluded the answering of factual questions for the period of earlier administrations. Her reply was that:

> the convention is that the Government of the day do not normally disclose information relating to the period of a former administration *and not published during that period*.[112] (Emphasis added.)

It was seen earlier that Benn, when minister, had indeed made a statement on his satisfaction with the correct arrangements for plutonium export and use. It is probable that the Prime Minister chose not to refer to this because Benn, when out of office, disowned his own ministerial statement!

The denouement in the pursuit of the truth of this matter came perhaps on 10 February 1987 when Tam Dalyell, both a staunch supporter of nuclear power and staunch critic of Mrs Thatcher over what he maintains are her lies to Parliament over the sinking of the Argentinian ship the *General Belgrano* during the Falklands War, the Westland helicopter affair and the American bombing of Libya, asked the Prime Minister if she would specify the types of question which she was prepared to answer which related to the periods of previous administrations. The reply read:

> As a *general rule* I do not answer questions relating to the policies and decisions of previous Administrations, for which I have no ministerial responsibility.[113] (Emphasis added.)

As the rule is general not inclusive, this permits the Prime Minister to pick and choose which issues suit her for prime ministerial commentary and which not. Thus statements have been made on Sir Anthony Blunt, Sir Maurice Oldfield and Sir Roger Hollis on matters of the state security services, but the issue of plutonium remains debarred. When Chris Smith followed up his questions of 15 and 18 April 1986 on plutonium, he was told on 27 February 1987 by the Prime Minister that she had:

> nothing to add to the [previous] replies.[114]

This refusal to make any further disclosure should be understood in the context of the official policy on information disclosure on plutonium enunciated by the Department of Energy on 12 December 1986. When asked if the Secretary of State had any plans to review the policy of commitment to openness in relation to information on the production, reprocessing, storage and use of civil plutonium arising from civil Magnox reactors, Goodlad replied:

> It is my policy to be as open as possible on these matters.[115]

Conclusion

The secrecy surrounding matters of national defence traditional in the United Kingdom was greatly exacerbated by the advent of nuclear weapons. As is well known, nuclear weapons decision-making in Britain since the war has, on a number of occasions, flouted accepted constitutional practice and there has been a continuing and by no means completed struggle to subject it to parliamentary accountability.

The links between nuclear energy and nuclear weapons in Britain, not a public issue in the 1950s, were allowed, if not encouraged, by successive governments to become the subject of a widespread public misconception. The myth that nuclear power and nuclear bombs were quite distinct developments in their respective countries, so convenient for successive British and American governments, flourished even in normally well informed or concerned circles, in spite of the fact that evidence to the contrary did exist in archives in the public domain. The British Government, and the Central Electricity Generating Board, could not readily contemplate the dispelling of that myth. Faced with nuclear issues as a controversial issue on the national agenda, the Government was driven to constitutionally questionable evasions and obfuscations, and the Central Electricity Generating Board to unsustainable assertions and downright lies.

Something of the aura of unquestionability, the hint of overriding but unmentionable national interests, so powerful an asset to defence decision makers, has accrued to national nuclear energy policy makers. The full extent to which military considerations have distorted this aspect of British energy policy can only be guessed at. But that significant distortion has occurred, both within the nuclear power programme, in terms of reactor choice, and between nuclear and other forms of electricity generation, now seems beyond doubt.

Notes

1. R. C. K. Ensor, *England 1870–1914*, Oxford University Press, 1936, pp. 394–5.
2. T. Benn, *Guardian*, 21 November 1983. In a personal communication with the present author, dated 3 January 1984, Mr Benn stated that he had 'done his best' from his own papers and tape recordings but that 'by definition you cannot prove what is not in the public domain and that means that my evidence has to rest on my own experience and credibility.'
3. The Combined Policy Committee for Uranium Procurement in Britain's African colonies continued in being.

4. See especially the 4-volume RAND Corporation study supported by the American National Science Foundation: R. Perry et al., *Development and Commercialisation of the Light Water Reactor, 1946–1976*, R-2180-NSF, June 1977; W. Allen, *Nuclear Reactors for Generating Electricity: U.S. Development from 1946 to 1963*, R-2116-NSF, June 1977; E. Rolph, *Regulations of Nuclear Power: the Case of the Light Water Reactor*, R-2104-NSF, June 1977; A. Gandara, *Electricity Utility Decision Making and the Nuclear Option*, R-2148-NSF, June 1977. See also I. C. Bupp and J. C. Derian, *The Failed Promise of Nuclear Power: the Story of Light Water*, Basic Books, Harper Colophon, New York, 1981.
5. Perry, see note 4. See also P. Pringle and J. Spigelman, *The Nuclear Barons*, Michael Joseph, London 1983, especially chapter 10.
6. John Simpson, *The Independent Nuclear State: the United States, Britain and the Military Atom*, Macmillan, London, 1983, pp. 59–60.
7. S. Rippon, 'The historical place of the PWR in energy supply', *Proceedings* of Issues in the Sizewell B Inquiry, volume 1. Centre for Energy Studies, London, 1983.
8. S. Rippon, 'History of the PWR', *Energy Policy*, vol. 12, no. 3, September 1984, p. 259.
9. Allen, see note 4, pp. 15–9.
10. R. H. Olby, *The Case for Nuclear Power Examined*, 1979, p. 53, Report No. 1979: 5. Scientific Expertise and the Public, Oslo.
11. A. H. Weaving, 'The development of the pressurised water reactor', *Nuclear Energy*, vol. 18, no. 2, April 1979, pp. 101–15.
12. Simpson, see note 6, pp. 113–4.
13. Sir John Cockroft, *Financial Times*, 9 April 1956.
14. Weaving, see note 11, p. 102; Rippon, see note 8, p. 259.
15. O. Townsend, 'The atomic power program in the United States', in *Atoms for Power*, P. C. Jessup (ed.), Columbia University Press/The American Assembly, New York, 1957, pp. 35–79.
16. See Allen, see note 4, pp. 28–34.
17. Bupp and Derian, see note 4, pp. 32–3.
18. Allen, see note 4, pp. 34–6.
19. Ibid., pp. 39–53.
20. *Financial Times*, 9 April 1956.
21. John Baylis, *Anglo-American Defence Relations, 1939–1980*, Macmillan, London, 1981.
22. Simpson, see note 6, p. 116.
23. Duncan Campbell, *The Unsinkable Aircraft Carrier*, Michael Joseph, London, 1984, p. 4.
24. Simpson, see note 6, p. 117.
25. Ibid., p. 118.
26. *Keesings Contemporary Archives*, 6–13 October 1956, p. 15133.
27. Simpson, see note 6, p. 119; Baylis, see note 21, p. 54.
28. David Fishlock, 'How the Navy has reared its nuclear family', *Financial Times*, 29 January 1982.
29. Simpson, see note 6, p. 122.
30. Townsend, see note 15, p. 120.
31. L. Hannah, *Engineers, Managers and Politicians*, Macmillan, London, 1982, p. 175.
32. Ibid., p. 181.
33. Although the phrase 'atomic virility' well encompasses the mood of the times, Derek Robinson's *Just Testing*, Collins, London, 1985, making use of interviews with participants in the British H-bomb test programme at Christmas Island and Malden Island in 1957, shows how 'virility' itself was damaged and questioned in retrospect.

34. K. Hubbard and M. Simmons, *Operation Grapple: Testing Britain's First H-Bomb*, Ian Allen, London, 1985.
35. Simpson, see note 6; R. V. Hesketh, *Nuclear Power UK, Nuclear Weapons USA: a Proof of Evidence on Behalf of CND to the Sizewell 'B' Public Inquiry*, September 1984.
36. *Keesings Contemporary Archives*, 16–23 November 1957, p. 15859; *The Times*, 4 November 1957.
37. *Keesings Contemporary Archives*, 26 October–2 November 1957, p. 15823.
38. *Department of State Bulletin*, 11 November 1957, p. 740.
39. *Guardian*, 2 May 1986.
40. *Daily Telegraph*, 1 January 1988.
41. Ibid.
42. The hearings were held on 29–31 January; 4–5, 27 February; 5, 26–28 March; 17 April; and 28 May 1958. Simpson, see note 6, estimated that about one third of the evidence given was never printed because of national security considerations.
43. Simpson, see note 6; Hesketh, see note 35.
44. *Atom*, no. 287, September 1980, p. 222. These are plutonium production figures for reactors that are on a civil, i.e. electricity-optimised, cycle; the plutonium capacity for reactors on military cycles is a national secret in each country that produces military plutonium.
45. Hearings before the Subcommittee on Agreements for Cooperation of the Joint Committee on Atomic Energy, 85th Congress, 2nd session, *Amending the Atomic Energy Act of 1954 — Exchange of Military Information and Material with Allies*, Washington, DC, 1958, p. 204.
46. Ibid., p. 132.
47. Ibid., p. 216.
48. Ibid., pp. 216–8.
49. Simpson, see note 6, p. 132.
50. Ibid., pp. 135–6.
51. On 17 June 1958 the Ministry of Defence, which did not own nor operate the CEGB nuclear reactors, issued a statement which concerned 'the production of plutonium *for weapons* in the new power station programme as an insurance against *defence needs*' (emphases added). This was followed by parliamentary questions: *Commons Hansard*, 24 June 1958, col. 234–6; 30 June 1958, col. 857–8; 24 July 1958, col. 247–8; in the last entry, in response to the suggestion that it was a disgusting imposition on what was termed a primarily peaceful programme in nuclear energy to 'modify the Magnox programme for military needs' the Minister (Reginald Maudling) replied that, 'the only imposition on the country would have arisen if the Government had met our defence requirements for plutonium by means far more expensive than those proposed'.
52. Simpson, see note 6, p. 140.
53. *Reflections on Britain's Nuclear History: a Conversation with Lord Hinton by David Lowry*, 19 January 1983, ERG-048, Energy Research Group, Open University, Milton Keynes, September 1984, pp. 21–2.
54. Hannah, see note 31, p. 235.
55. *Nuclear Engineering*, vol. III, no. 26, May 1958.
56. In fact the Anglo-American Mutual Defense Agreement reduced the benefit of recycling plutonium, as the plutonium credit had reduced the estimated unit cost of electricity by one third in 1955, but it only benefitted by one twelfth or one fifteenth by 1958: Hannah, see note 31, p. 230.
57. Hannah, see note 31, p. 236.
58. Ibid., p. 239.
59. *Nuclear Engineering*, vol. III, no. 29, August 1958.

60. Simpson, see note 6, p. 146; *Commons Hansard*, 11 July 1958, col. 779–94 and 1 August 1958, col. 228–9.
61. *Commons Hansard*, 7 May 1959, col. 63–4: statement by the Prime Minister.
62. *The Times*, 6 May 1959, reported that the Admiralty had announced plans by 8 different companies to build reactors for Britain's submarines.
63. *Daily Telegraph*, 8 May 1959.
64. *The Times*, 8 May 1959.
65. Ibid.
66. On 4 May 1959 the President also endorsed agreements on mutual defence cooperation with West Germany; on 26 May agreements with the Netherlands were transmitted to Congress. Each was to do with American deployment in the respective countries. Public Papers of President Eisenhower, Office of the Federal Register, National Archives and Records Service, General Services Administration, items 119, 120, pp. 421–422, 1959.
67. David Lowry, *Nuclear Powers: an Assessment of Nuclear Decision Making, 1932–1979, with special reference to the Anglo-American atomic relationship*, doctoral thesis, Energy Research Group, Open University, Milton Keynes, November 1986.
68. *Commons Hansard*, 14 May 1959, col. 1426–27.
69. Ibid., 21 November 1960, col. 774–818.
70. Ibid., 27 March 1963, col. 1168.
71. J. Simpson, 'The Anglo-American nuclear relationship and its implications for the choice of a possible successor to Polaris', in House of Commons Expenditure Committee, *The Future of the United Kingdom's Nuclear Weapons Policy*, 6th Report, HC Paper 348, 1978–79, p. 223.
72. Hinton, see note 53, pp. 24–8. A ministerial statement, *Commons Hansard*, 22 June 1959, col. 849, claimed that only Hinkley Point 'A' Magnox station was modified for potential military use. But see also Barnham *et al.*, note 83 below and Richard Norton-Taylor, *Guardian*, 15 October 1987.
73. 'Poisoning the peaceful atom', *New York Times*, 30 September 1981.
74. 'Civil plutonium not used for arms', *Financial Times*, 15 January 1983.
75. Hinton, see note 53.
76. 'US wants to buy British atom fuel', *Guardian*, 12 October 1981; 'Weapon fuel denied', *Guardian*, 13 October 1981; 'Britain's atom fuel to make up US shortfall', *Guardian*, 13 October 1981.
77. Martin Ryle, 'The use of exported plutonium', *The Times*, 17 October 1981.
78. R. V. Hesketh, 'The use of exported plutonium', *The Times*, 30 October 1981.
79. S. Bhatia, 'Electricity Board critic is dismissed, *Observer*, 12 June 1983; Richard Norton-Taylor, 'Scientist claims sack was for exposing arms link', *Guardian*, 13 June 1983; David Lowry, 'The dangers of spilling the beans about plutonium', *Guardian*, 15 June 1983; Peter Pringle, 'CEGB sacks man who exposed plutonium link', *New Statesman*, 17 June 1983; David Lowry and R. V. Hesketh, 'Electricity Board fires its nuclear critic', *Observer*, 19 June 1983; S. Alderson, 'Behind the nuclear power station's smokescreen', *Guardian*, 20 June 1983; 'Don't fire the messenger: plutonium dissident loses appeal', *New Scientist*, 23 June 1983; Richard Norton-Taylor, 'Nuclear protest scientist loses sacking appeal', *Guardian*, 23 June 1983; J. Armstrong, 'In the heat of the conflict', *New Society*, 23 June 1983; C. Sweet, 'Energy, Sir Walter and the power of free speech', *Guardian*, 24 June 1983; J. W. Baker, 'Plutonium pledge', *Observer*, 26 June 1983; A. Reddish *et al.*, 'Nuclear scientist's dismissal', *Times Higher Education Supplement*, 1 July 1983; R. V. Hesketh and Martin Ryle, 'Nuclear answers needed', *Observer*, 3 July 1983; T. Broom, 'CEGB answer back', *New Scientist*, 7 July 1983; J. Anderson, 'Plain facts about a scientist', *Guardian*, 7 July 1983; D. Dickson, 'Firing spotlights plutonium exports', *Science*, 13 July 1983; C. Sweet and R. V. Hesketh, 'Arguments the CEGB

dismissed too easily', *Guardian*, 18 July 1983; P. Bateson *et al.*, 'Sacking of Dr Hesketh', *The Times*, 19 July 1983; D. M. Woodroffe, 'Dr Hesketh's dismissal', *The Times*, 27 July 1983; R. V. Hesketh, 'CEGB fired', *New Scientist*, 28 July 1983; Richard Norton-Taylor, 'Demotion claim by scientist's colleagues', *Guardian*, 12 December 1983; R. Edwards, 'Union leaders attacked over Hesketh sacking', *New Statesman*, 16–23 December 1983.
80. D. Fishlock, 'The plutonium hot potato', *Financial Times*, 27 October 1981.
81. N. Dombey, 'Fuelling suspicion', *Guardian*, 3 December 1981.
82. B. Groom, 'Power engineers warn on plutonium exports to US', *Financial Times*, 13 April 1982.
83. K. W. J. Barnham, D. Hart, J. Nelson and R. A. Stevens, 'Production and destination of British civil plutonium', *Nature*, vol. 317, 19 September 1985, pp. 213–7.
84. Ibid., p. 216.
85. Published by Pluto Press, 1982.
86. 'Sellafield and the bomb', *TV Eye*, Thames Television.
87. Hinton, see note 53, p. 25.
88. Ibid., p. 26.
89. Roger Williams, *The Nuclear Power Decisions*, Croom Helm, London, 1980, p. 101.
90. Hinton, see note 53, p. 37.
91. Ibid., pp. 37–8.
92. P. O'Brien, 'Plutonium for 150 bombs disappears', *Ynni*, September/October 1982, p. 3; 'Wylfa built for nuclear weapons', *Western Mail*, 9 October 1984; S. Neville, 'Wylfa weapons link', *Western Mail*, 8 October 1984; M. Jones, 'Links with N-bombs denied by Wylfa', *Holyhead & Anglesey Mail*, 1 November 1984.
93. Lowry, see note 67.
94. H. Macmillan, *At the End of the Day*, Macmillan, 1973, p. 121; quoted by Simpson, see note 6, p. 158.
95. Simpson, see note 6, p. 159.
96. A. Gorz, *Ecology as Politics*, South End Press, Boston, 1980, pp. 104–5.
97. *Commons Hansard*, 12 January 1976, col. 1–2.
98. Republished in: *The Sizewell Syndrome; Nuclear Weapons, Nuclear Power and Public Policy*, END Paper 7, Spring 1984.
99. *Commons Hansard*, 3 December 1984, col. 24.
100. Ibid., 15 January 1985, col. 85–6.
101. Ibid., 13 May 1986, col. 612.
102. Ibid., 23 February 1987, col. 53.
103. *Guardian*, 20 January 1983.
104. A point made by the present author in 'Why Mr Lawson can't obscure links in this nuclear chain', *Guardian*, 24 January 1983.
105. *Commons Hansard*, 28 January 1987, col. 259.
106. R. Edwards, *Nuclear Power, Nuclear Weapons: the Deadly Connection*, CND Publications, 2nd edition, 1987; A. Turner (rapporteur), Opinion for the Committee on Energy, Research and Technology on the motion for a resolution (Doc. 2-1250 (84) *On the Contravention by the UK Government of the Euratom Treaty*, P. E. 102 190/fin. Committee on Legal Affairs and Citizens' Rights, 20 December 1985; L. Smith (rapporteur), *Draft Report on the Contravention by the United Kingdom Government of the Euratom Treaty*, (Document 1250/84), P. E. 106, 724, Committee on Energy, Research and Technology, 11 June 1986; *Reply by Commissioner Mosar to Mr Poniatowski's letter of 27 February 1987 containing a series of questions on safeguards at Sellafield*, P. E. 112, 357, Directorate-General for Committees and Delegations, European Parliament, 26 March 1987.

107. David Lowry, *A New Nuclear Glasnost?*, paper presented to the Conference on Nuclear or Non-nuclear Futures at the Polytechnic of the South Bank, London, April 1987. Reprinted in the proceedings.
108. *Commons Hansard*, 4 February 1983, col. 206.
109. Ibid., 15 April 1986, col. 330.
110. Ibid., 18 April 1986, col. 525.
111. Ibid., 8 May 1986, col. 217.
112. Ibid., 14 July 1986, col. 320.
113. Ibid., 10 February 1987, col. 161.
114. Ibid., 27 February 1987, col. 429.
115. Ibid., 12 December 1986, col. 245–6.

Appendix
Source: Hansard 14 May 1987

United Kingdom–United States of America (Treaties)

Mrs. Clwyd asked the Secretary for Defence if he will list all treaties, agreements, memoranda of understanding and other formal arrangements currently in force between the United States of America and the United Kingdom, covering the collaborative research, development and production of defence equipment and British participation in United States defence programmes.

Mr. Archie Hamilton [*pursuant to his reply*, 1 May 1987, c. *300*]: The major arrangements currently in effect between the Governments of the United Kingdom and the United States covering collaborative research development and production of defence equipment are as follows. We have tried to make these lists as comprehensive as possible. However, there are numerous relatively minor arrangements, particularly in research and information exchange, of which it would be disproportionately costly to attempt to compile a comprehensive schedule. The lists do not include arrangements of a classified nature; nor do they cover supplies of British defence equipment to the United States under purely commercial arrangements.

General

Mutual Defense Assistance Agreement between the United States of America and the United Kingdom of Great Britain and Northern Ireland, signed 27 January 1950 (Cmnd. 7894).

Agreement between the Government of the United Kingdom of Great Britain and Northern Ireland and the Government of the United States of America to Facilitate the Exchange of Patents and Technical Information for Defense Purposes, dated 19 January 1953 (Cmnd. 8757).

Mutual Defense Assistance Agreement between the United States of America and the United Kingdom of Great Britain and Northern Ireland, signed 19 January 1953.

United States of America and the United Kingdom of Great Britain and Northern Ireland General Security Agreement, signed 14 April 1961.

Agreement Relating to Defense: Weapons Production Program between the United States of America and the United Kingdom of Great Britain and Northern Ireland, dated 29 June 1962 (Cmnd. 1863).

United States of America and the United Kingdom of Great Britain and Northern Ireland Memorandum of Understanding Arrangements for Joint Military Development, May 1963.

The Basic Standardization Agreement Among the Armies of the United States of America, the United Kingdom of Great Britain and Northern Ireland, Canada, and Australia: 1 October 1964.

Memorandum of Understanding between the Government of the United States and the Government of the United Kingdom of Great Britain and Northern Ireland relating to the Principles Governing Cooperation in Research and Development, Production, Procurement and Logistic Support of Defence Equipment dated December 1985.

5

Policy Making in Civil and Military Electronics: the Limits of Pragmatism

MARGARET BLUNDEN, CHRIS BISSELL
and JOHN MONK

Introduction

Electronics and information technology have for more than a quarter of a century assumed a dominant role in both military and industrial development. The health of a country's electronics industry is increasingly correlated with its general economic prosperity. Levels of investment in new electronics, both directly and as an enabling technology to promote industry efficiency, are consequently now widely seen as critical to the economic future of industrialised countries. At the same time, electronics has becoming a crucial military technology. Sophisticated, costly hardware and software have already transformed weapons systems, nuclear and conventional, and most major proposals for strategic change, from both right and left of the political spectrum, are now dependent on the capabilities of electronic sub-systems.

These developments pose problems for both military and civil areas, and raise serious questions about the connections between them. The choices opened up by sophisticated military electronics are creating a more unpredictable military environment and are placing great strain on the defence budgets of all the militarily advanced countries. For instance, the costs of military aircraft have been increasing by a factor of four every ten years, reflecting in particular the growing trend towards costly and sophisticated avionics. It has been noted wryly that, 'If the trend continues, in the year 2054 the entire defence budget [of the United States] will purchase just one aircraft'.[1] One of the aims of this chapter is to explore this phenomenon of increasingly complex military electronics, and point to its, as yet little acknowledged, budgetary consequences.

As civil electronics, for its part, becomes more complex, more dependent on high levels of research and development, and more internationally competitive, the question of how military programmes affect the civil electronics industry becomes more crucial. Is it possible to generate commitment to high levels of investment in science and technology without going through the military route, with its strong imperative and powerful sources of appeal? Or are high levels of investment in military research and development both a necessary and an efficient way of promoting the civil electronics industry? The politically contentious issue of the effect of military electronics on civil electronics is particularly significant in those Western countries—notably the United States, the United Kingdom and France—where military research and development represents a large and growing proportion of total national research and development. The issue is particularly acute in Britain, whose civil electronics industry has been steadily declining in international competitiveness for the past 20 years.

This chapter looks at the different national strategies towards civil and military electronics of the United States, Japan, France and the United Kingdom and explores the policy dilemmas, posed in a particularly acute form in Britain, arising from apparently conflicting military and civil interests. How best to reconcile civil and military interests in electronics—how to jointly optimise these two systems so often separated politically and bureaucratically—is as important a question for defence as for industry, since national security in its broadest sense must depend on the existence of a flourishing economic base.

The research which has been done on the British electronics industry is by no means exhaustive; but the conclusions which flow from it point in rather different directions from the pragmatic and *ad hoc* policies which have actually been pursued.

Military electronics: trends and policy dilemmas

Although the unit costs of civil electronics have been falling dramatically, the electronics component, hardware and software, of modern military systems is becoming more expensive and is disproportionately responsible for the escalating unit costs of modern weapons.[2] It is in electronics that the cost-reducing impact of technological change on civilian production, and its cost-amplifying effect on military production, is most marked. This section explores the phenomenon of rising costs in military electronics, with its potential to play havoc with inelastic defence budgets.

The percentage of defence budgets taken up by equipment is rising in the militarily advanced countries. In the United Kingdom, which devotes a larger share of defence spending to equipment than any other NATO member, it rose from 41 per cent in 1979 to 45 per cent in 1987.[3]

This increasing percentage reflects long-standing cost escalation in the major weapons systems of all three services, increasing from one generation to the next by 6 to 10 per cent faster than the general rate of inflation, and causing longer gaps between the replacement of each generation of new equipment. The difference in cost between the F-104G Starfighter of 1965 compared with the Tornado multi-role combat aircraft of 1980 represented an increase, after allowing for inflation, of 10.5 per cent per year. The Leopard 2 tank of 1980 represented a cost increase of 6 per cent per year compared with the M47 tank of 1956. The type 42 destroyer of 1975 represented a cost increase of 6 per cent per year compared with the *Defender* destroyer of 1952.[4]

Rising costs are mostly due to the increased sophistication of weapons systems, sophistication gained mainly by installing more electronic subsystems. Modern military systems depend, and will increasingly depend, on electronics for their performance. Electronics plays a vital role, not only in modern weapons systems, nuclear and conventional, but in command and control networks, in intelligence gathering and interpretation, in verification of arms-control agreements, in military training (such as simulation exercises) and in military management at all levels. The proportion of defence equipment budgets being spent on electronics has risen rapidly. In the United Kingdom, for instance, it was estimated that 36.5 per cent of the 1985–86 defence procurement budget would be spent on electronics compared with 28 per cent five years earlier,[5] and Ministry of Defence spending on electronics is expected to continue to rise, in spite of planned level funding in the defence budget. Total British defence expenditure on electronics amounted to £1.4 billion in sales in 1984 and this is expected to rise to £2.3 billion in 1994.[6] The *proportion* of the total British defence procurement budget currently taken by electronics is comparable to that of the United States (with electronics accounting for an estimated 35 per cent of the total procurement budget in 1987), although, of course, in terms of total spending there is an enormous gap between the United States and the Soviet Union on the one hand, and the medium-sized powers, such as the French and the United Kingdom, on the other.

In terms of individual weapons systems, the average electronics proportion by value was in 1985 approximately 33 per cent for aircraft, 45 per cent for missiles and 66 per cent for satellites.[7] For the United States worldwide military command and control system, a vast undertaking which the smaller powers could not hope to emulate, the proportion of electronics is 88 per cent, and this is comparable with the proportion of electronics in the Strategic Defense Initiative (SDI) programme. These are average figures: the cost of hardware and software for fighter aircraft has risen particularly quickly—from 2 per cent of the cost of the F-4 Phantom in the 1960s to 43 per cent of the current American F-18 programme[8]—over a period when computer costs in general have fallen by several orders of magnitude. The avionics of a modern fighter aircraft may amount to 60 per cent of its cost. Software costs alone may make up half the costs of an advanced weapon system. The Royal Navy's new Type-23 frigate, for instance, now under construction by Yarrow Shipbuilders may well have 200 microcomputers on board to control its weapon and sensor systems. The designing and writing of the software for these systems could amount to half the total cost of getting the new frigate into service.[9] Rapidly rising total software costs for military systems represent vast sums for a superpower like the United States, where such costs for major weapon systems were $4 billion in 1980, $10 billion in 1986 and are predicted to be $30 billion per year by the early 1990s.[10]

There are good reasons why the trend towards the increasing importance of electronics in military systems is unlikely to be arrested, let alone reversed. Although there are people on both the right and the left of the political spectrum who want to get away from high technology defence policies, all nuclear strategies and many alternative conventional strategies are highly dependent on electronics. For instance, the strategy of 'follow on forces attack' (FOFA), espoused by General Bernard Rogers during his time as Supreme Allied Commander, Europe, with the objective of raising the nuclear threshold, relied heavily on electronic advances in missile guidance to attack enemy forces up to 400 kilometres behind the enemy lines. Similarly, some radical suggestions for defending Europe by non-provocative as well as non-nuclear means depend on the electronic sophistication of electro-optical sensors, 'smart' anti-tank mines, computerised battlefield information systems and short-range precision-guided missiles.[11] The agreed elimination of intermediate range nuclear forces in Europe is likely to increase NATO's emphasis on technical superiority in conventional weapons, which in practice means more sophisticated electronics.

Electronics not only lies at the heart of modern nuclear strategies such as the Reagan administration's hope to prevail even after a prolonged nuclear war but it is simultaneously transforming conventional warfare. Modern conventional warfare relies increasingly on missiles, as recent conflicts in the Falklands and the Middle East have demonstrated. Micro-electronics is by far the most important technology underpinning the development of more intelligent missiles. Major Western armies and the Soviet army have been deploying 'smart' conventional missiles—that is, radar or infra-red homing missiles—since the 1960s. But smart missiles, dependent as they are on an operator to identify and designate a target for them, are not smart in terms of modern electronic potential. An electronics-based research programme now under way in the United States has the objective of realising what are called 'brilliant' missiles, that is, missiles with the capability of distinguishing targets such as tanks from non-targets such as rocks, of reporting back to an operator (who might be some 100 miles away) the kinds of target identified, of receiving from the operator, by voice, instructions about the priority to be given to the different kinds of target identified, and of then proceeding to attack and destroy them. The research programme for brilliant missiles involves five critical branches of electronics technology: high resolution sensors; artificial intelligence; algorithms for pattern recognition; microprocessors; and precision guidance.[12] Even if brilliant missiles in this form are not realized, the heavy investment in this research programme, as in the SDI programme, will accelerate developments in the field, with potential military applications as yet unforeseen.

Potentially, advances in electronics may make possible the battlefield commander's dream (or nightmare): the automated battlefield in which a commander is able to see the field of conflict in real time and to send out orders simultaneously. General William Westmoreland, ex-Chief of Staff of the United States Army, has predictd that, 'on the battlefield of the future, enemy forces will be located, tracked and targeted almost instantaneously through the use of data links, computer-assisted intelligence evaluation, and automated fire control.'[13] Whether such a battlefield would be too complex and unpredictable for any commander to control is an open question, beyond the scope of this chapter. Whether this technology, too, can be achieved in this form is again not yet clear. What is clear is that advances in electronics already achieved have made conventional warfare much more destructive than ever before. One salvo from the Multiple Launch Rocket System, whose terminally guided warheads are being collaboratively developed by the United States, the United Kingdom, France and West Germany, for instance, may cause as many casualties in the short term as a small nuclear weapon.[14] Consequently, the modern battlefield presents an increasingly hostile technological environment to soldiers and this in turn has stimulated developments of robots and unmanned vehicles, even more dependent on advanced electronics.[15]

The process by which more electronics in missiles has led to more electronics for military robots is characteristic of the self-stimulating proliferation of military electronics. Electronic measures are leading to electronic counter-measures and electronic counter-counter-measures, in a seemingly unending spiral. For instance, the development of radar was followed by radar-jamming techniques, such as continuous-wave noise transmitters with greater band-width than the radar receiver and with uniform power over the band-width. This led to counter-radar-jamming techniques, which evaluated the level of continuous-wave noise and cancelled its effects, and to missiles that could home in on jamming transmitters. In reply, pulsed repeater jammers, which cannot be countered by conventional anti-jam techniques, were devised. Then radar designers developed new signal-processing algorithms to counter pulsed-jamming, based on the receivers' abilities to recognise their own pulses.[16] Similarly, many companies in the United States, the United Kingdom and France are working on meeting the threat posed to combat aircraft by laser target-designators, laser range-finders and laser-guided missiles, by developing laser warning-systems, often integrated with the radar warning-receiver and using the same display. Electronic counter-measures and electronic counter-counter-measures are intensely competitive areas, since these are thought to be two of the few areas of electro-technology in which the Soviet Union is level pegging with the United States.

Massive American investment in military electronics R & D is designed to ensure that the country's technological superiority over the Sovet Union is maintained. For instance, the six-year Very High Speed Integrated Circuit programme is aimed at developing silicon chips fast enough and reliable enough to provide the speed and operation required for real-time monitoring and control. New projects include sonar-identification buoys, jam-resistant radar systems and electro-optical signal processors.[17] Such projects set the pace, not only for the Soviet Union, but for major allies of the USA such as the United Kingdom and France. They help to sustain a high rate of change and obsolescence in military equipment, and continuing pressures to maintain if not to increase levels of spending on military R & D.

The budgetary pressures generated by high rates of change and obsolescence cannot easily be mitigated by the kinds of reduced unit cost seen in parts of the civil electronics sector. For one thing, the increasing unit costs of military electronics and the decreasing unit costs of civil electronics reflect very different kinds of specification. The military systems of the major military equipment manufacturing nations are designed to be rugged enough to operate in most of the natural environments found on the planet. (This reflects primarily political aspirations to play a world role, but has the commercial advantage of producing weapons systems suitable for worldwide export.) Military equipment, including electronic systems, is designed to operate in extremes of temperature and humidity, and operation over these ranges adds to the cost. Although only parts of weapons systems might normally be expected to suffer extremes of shock and vibration, other parts are designed to similar specifications in case of possible battlefield incidents. The standard military temperature range for semiconductor devices is $-55°C$ to $+125°C$, which is beyond the commercial range of 0°C to +70°C or the industrial range of −40°C to +85°C. Similarly, military requirements may demand distinctive packaging techniques. For example, military components are often packaged in ceramics whereas commercial components are packaged in plastic. The military components, packaged as they are in intrinsically more expensive materials, do not benefit from the accumulated knowledge and economies gained in large-scale, commercial production. For example, the AMD2 6LS20 is a chip for driving signals along cables. In a plastic package to commercial specification it costs £2.18; in an hermetic package to commercial specification it costs £3.39; and to military specification, which is available only in an hermetic package it costs £7.57, over three times more than the cost of the commercial equivalent in a plastic package.[18]

Military systems are also likely to be specified to higher degrees of reliability than most domestic or even industrial systems, except where safety is a major factor. Space-borne electronic systems, for instance, must not only be small and light, able to withstand the rigours of being launched into space, extremes of temperature and the effects of cosmic radiation; they must also be highly reliable, as repairs can rarely be considered and it is difficult to instal replacements. Growing insecurity about NATO's supply lines and remote spares depots and the Reagan administration's aspiration to survive any prolonged nuclear war, have shifted American technological priorities towards highly reliable weapons systems that could withstand lengthy battles.

As a result, modern military systems are heavily dependent on the careful testing of each of the thousands of small components which go to make them up. The consequences of the inadequate testing of microchips—such as that allegedly practised by Texas Instruments, a major supplier to American defence contractors[19]—may be far-reaching and incalculable. Self-repairing, fault-tolerant and self-testing electronic systems are feasible, although they naturally contain more electronics than the systems on which they are based and are disproportionately costly. Fault tolerance is achieved by adding more components or by repeating the signals that flow in the system. Repetition of signals can prevent some kinds of failure, but the extra signals reduce the capacity of the system and faster components are needed at additional cost to restore the performance to its required level. Consequently, triple redundancy, for instance, more than triples the cost. The American F-16C combat aircraft, which is expected to enter service in 1988, uses a quadruply redundant computer, at more than quadruple cost, to achieve fault tolerance in its critical flight controls. Increasing sophistication may be able to increase reliability: the United States Department of Defense claimed in 1982 that the mean time between maintenance actions for an electronic system, such as radar, was doubling as later generations of electronics were being introduced.[20] However, even this has its unexpected costs: where complex systems do fail, only highly specialised personnel with specialised equipment are likely to be able to diagnose and repair the fault.

An important and distinctive cost factor in military electronics is their potential vulnerability to the electromagnetic pulse (EMP), a sharp, intense pulse of radio waves generated by nuclear explosions, and to the effects of nuclear radiation. A single, high-altitude, megaton explosion could, for instance, knock out most telephone communications, leaving satellites, whose electronics are themselves highly vulnerable to the effects of nuclear weapons, as the only long-distance communications link. Most military equipment now being designed for military use has to meet specifications for hardening against radiation and the EMP: special component design, circuit design and special technologies are needed to allow circuits to continue to operate when exposed to abnormal levels of radiation. Silicon on sapphire (SOS) is one of the materials that is able to withstand high bursts of radiation. The circuits are built on a substrate of sapphire, which is an insulator. It is difficult to dislodge electrons from an insulator and even when dislodged, they do not flood to the surface where the circuit is. Not only are silicon on sapphire chips not subject to upset from cosmic radiation in outer space, they may also be faster and consume less power than alternative chips. But, unlike optical communications which are both resistant to EMP and have civil applications, silicon on sapphire is not a material used in industrial or domestic electronics: it is too expensive. This means that there is little possibility of cutting defence costs by taking advantage of commercial economies of scale, or, indeed, of securing economic benefits by means of spin-off from military to civil applications.[21]

The extent to which military electronics may be made EMP resistant is limited in the current state of the art: there is no known technology by which to harden very large-scale integrated and very high-speed integrated circuits to the levels required to make their use in satellites invulnerable, and capability has to be traded against vulnerability even at lower levels of sophistication. But the emphasis on hardening wherever possible contributes to the big differences in costs between military and civil equipment serving broadly the same functions. What is more, the emphasis on hardening complicates the question of procuring for military use electronics equipment designed for the civil market, a practice otherwise desirable in terms of speed, economies of scale and benefits to the civil electronics industry at large, and especially to small contractors.

Research, development and production of major weapons systems is a long process: it may take as long as 15 years from conception to entry into service. The desire to update weapons, as new electronic technologies are developed, has tempted defence bureaucracies constantly to redefine projects, especially in the case of cost-plus contracts, so as to incorporate, for instance, new developments in computing. But this often means that the contractor has to begin anew, and may lead, as in the case of the Nimrod airborne early-warning system, to additional costs and time overruns. The dilemma posed by the conflicting demands of invulnerability, speed and cost can only partly be avoided by buying equipment in quantity and replacing it if it is damaged. Most types of radiation and EMP damage is reduced for units not plugged in; but some damage, such as that from neutrons, occurs whether the equipment in question is connected to power or not and so could destroy spares as well.

The weapons procurement process is full of such cost dilemmas. Another example is the choice between procuring cheaper equipment (with more restricted uses and environmental tolerances) or more expensive, multi-role equipment (with wider environmental tolerances). The first option is cheaper because for each piece of equipment there is considerable scope for design trade-offs to reduce costs. The disadvantage is that equipment with restricted uses increases the range and the number of stocks that have to be carried, and the amount of training required to operate this larger number of systems. On the other hand, more expensive and versatile equipment can play a variety of roles in a variety of environments, and the range of stocks to be carried and training needs are reduced, thereby contributing towards reduction in mobilisation times. But equipment that withstands all the extreme conditions likely to be met will be more expensive and may need special technologies for its construction.

The tendency of defence bureaucracies to be sucked into over-elaboration and unnecessary expense—the process known as 'gold-plating'—and the problems so caused, were described in 1981–82 by Captain J. E. Moore, the editor of *Jane's Fighting Ships*:

> A basic frigate hull when provided with machinery for a speed of 25 knots, passive submarine detection systems and facilities for refuelling and re-arming an ASW helicopter could today be produced for about £30 million. This would be a ship designed specifically for the one wartime role, detecting and sinking submarines. As the docket circulates through the Ministry the speed is considered too low, a hangar is considered essential, active sonar is believed to be 'mandatory', she must be capable of both ESM and ECM, her visiting helicopter is not an adequate weapons platform so she must have a profusion of ASW weapons and, finally, she cannot be allowed to sea without adequate defence against aircraft and missiles. Thus, by the time the proposal reaches the Operational Requirements Committee for the second time ... the cost may be in the region of £70 to £75 million.[22]

What Captain Moore calls the 'Christmas Tree syndrome' and Mary Kaldor 'baroque technology',[23] that is, the ever-growing complexity and elaboration of electronics in military systems, is sometimes said to have led to weapons which not only cost too much, but which cannot be relied on to work, are complicated to maintain and operate, are of doubtful survivability, and present expensive and vulnerable targets. The wisdom and cost effectiveness of the current strong trend towards producing fewer aircraft and ships, each electronically more sophisticated and expensive, is itself disputed. As far as procurement goes, production runs are shorter, so that the benefits of economies of scale and learning curves are lost, while the heavy R & D costs can be spread over comparatively few units. Militarily, large advanced weapons systems present tempting and costly targets to cheaper, electronically sophisticated missiles. Many critics, for instance, regard airborne warning and control aircraft (AWACS), now deployed by the Americans and shortly to be purchased by the British Government, as an enormously expensive electronic white elephant, arguing that the aircraft would be a prime target in any war and might well be shot down within minutes.[24]

Be that as it may, the trend towards fewer, more expensive and elaborate weapons, reflecting cost increases for each new generation, continues. This contrasts with the mass production which has yielded such economies of scale in, for instance, consumer electronics. It seems beyond dispute that there will be increasing pressures on defence procurement budgets. Although one should never confuse expenditure on defence inputs with effectiveness of defence outputs, let alone equate increased spending with increased security, there are serious—and little acknowledged—policy difficulties which flow from this. In the United States, for instance, where even hawkish politicians think it unlikely that public opinion will support overall defence expenditure of more than about 6 per cent of the gross national product,[25] the impending painful choices among defence priorities have not yet fully been faced. The medium-sized powers, France and the United Kingdom, struggling on much more limited overall budgets to retain the advanced capabilities necessary to maintain a wide spectrum of defence commitments, have similarly yet to confront the full implications of their dilemma, although the clash of escalating weapons costs and budgetary constraints has already begun to undermine the defence consensus in France.[26] In Britain, the Government has seemed anxious to postpone for as long as possible the difficult choices which a defence review would involve, despite increasing disquiet about attempts to spread the defence budget too thinly.

The electronics industry: international comparisons

The pace of change in electronics and information technology has not only posed acute problems for defence budgets and defence policy, it

MAJOR COMPANIES IN THE DEFENCE ELECTRONICS, TELECOMMUNICATIONS AND ELECTRONIC SYSTEMS MARKETS*

Rank	Company	Group Sales in 1984/85 (£bn)	Rank	Company	Group Sales in 1984/85 (£bn)
1	IBM (US)	38.70	12	CGE (Fr)	6.57
2	AT & T (US)	27.96	13	Fujitsu (Jap)	5.31
3	Gen Electric Co (US)	23.54	**14**	**GEC (UK)**	**5.22**
4	Siemens AG (FRG)	14.05	15	Thomson (Fr)	5.13
5	Phillips (Neths)	12.93	16	Honeywell (US)	5.12
6	GTE (US)	12.25	17	Motorola (US)	4.66
7	Daimler-Benz (FRG)	11.79	18	AEG AG (FRG)	2.99
8	ITT (US)	9.41	19	L M Ericsson (Swed)	2.79
9	Westinghouse (US)	8.65	20	N Telecom (Can)	2.79
10	NEC (Jap)	7.67	**21**	**Plessey (UK)**	**1.42**
11	RCA (US)	7.31	22	Gould (US)	1.18

Source: *Monopolies and Mergers Commission*
* Home currencies converted to sterling using Dec 1984 exchange rates.

Source: The *Financial Times*, 7 August 1986

FIGURE 1

has also presented difficulties for strategic policy-making in civil electronics, on whose prosperity the military must ultimately depend. Leading advanced countries have adopted very different kinds of strategies to meet these challenges.

As may be seen from Figure 1, the United States dominates the world electronics industry, in terms of defence electronics, telecommunications and electronic systems. In the United States very high levels of investment in military electronics R & D for defence and space projects have *coincided* with a leading national role in miniaturisation, a flourishing civil electronics industry (especially in industrial rather than consumer electronics in which the Japanese are pre-eminent) and a high level of exports of defence electronics.

How far massive American defence spending has been responsible for promoting a healthy electronics industry is a contentious question, and the conclusions reached on it may say more about an author's views on the desirability of defence spending than about any hard evidence of an objective or quantifiable kind. The Congressional Joint Economic Committee concluded in 1985 that:

> The demand of the defense and aerospace markets pulled the [United States] industry along a specific trajectory and helped to prod the domestic sector into a position of market dominance and technological leadership. During the 1950s, while Europe and Japan pursued the development and mass production of germanium, transistor-based consumer electronic systems, the U.S. industry, prodded by U.S. military demands for miniaturisation, and devices of higher performance and reliability, became pre-eminent in silicon-based technology.[27]

Former Secretaries of Defense, and notably Harold Brown in 1980, have claimed that military expenditures are 'beneficial in the long term to the civilian economy.'[28]

Recently, however, a growing body of research has challenged this view, arguing that there has been little commercial application of devices developed for military purposes, incompatibility of organisational structures in commercial and military products, a poor record of innovation in military R & D and concentration of military effort on fields not relevant to civil development. Some economists cite evidence for a significant negative correlation between defence spending and general economic performance.[29]

What does seem clear is that, of the three countries (the United States, the United Kingdom and France) which Henry Ergas of the OECD [30] classified as mission-oriented—that is, primarily concerned economically with major projects of national significance, especially defence and/or space—the United States has been more successful than the United Kingdom in backing its military effort with solid success in commercially funded development. It has, in particular, been more successful than the United Kingdom or France in avoiding polarisation between a small number of high technology electronics firms, mainly oriented to sales in government markets, and the bulk of the industry which draws little benefit from public assistance to innovation. With the massive American government spending on defence and aerospace during the 1960s, funding was available for an enormous R & D effort with assured sales of digital integrated circuits at guaranteed high prices. It was government policy to support electronics developments in space and defence generously and consistently, and the financial and commercial climate was particularly favourable to small, entrepreneurial firms.

Specialist integrated circuit manufacturers, such as Fairchild and Texas Instruments, soon proved more capable of taking risky and creative steps to exploit the new situation than the older, vertically integrated electrical giants, even though the latter had been responsible for much of the early development work in the new fields. During the 1960s and early 1970s, many new names established themselves in American micro-electronics, often as a result of skilled personnel leaving larger firms to form their own companies. The 'Silicon Valley' phenomenon facilitated movement of personnel, flows of information and the technological and commercial expansion of many of the companies involved. In Europe, by contrast, there was a persistent tendency towards vertical integration which discouraged the establishment of small producers. Instead, specialised development took place within the existing, vertically integrated companies.

Whether American military and space programmes have actually stimulated innovation in electronics or not, the French believe that the

American programmes have done so. As R. Gilpin, the historian of modern French science policy, sees it, the French have quite consciously sought to duplicate the American experience of using nuclear weapons and space programmes to stimulate selected fields of scientific research and sectors of the economy, such as electronics. In his view of this controversial question:

> While investment in military and space programs may not be an efficient way, in terms of cost benefit analysis, to modernise a country's science and technology, in a democratic nation like France it is the only way to subsidise science and technology at the required level.[31]

However, in contrast to the United States, where it is at least arguable that massive military research and development contracts may have helped to launch the transistor and computer industries, the French military has done little to advance either technology. On the other hand, whereas the French spend less on military R & D, both in total and as a proportion of all R & D, than the British (see Figures 2 and 3), they have a larger share of the world arms market, with corresponding advantages to the domestic electronics industry. This reflects an export strategy of concentrating on a few, specialised areas of arms production, such as missiles and fighter aircraft, and a foreign

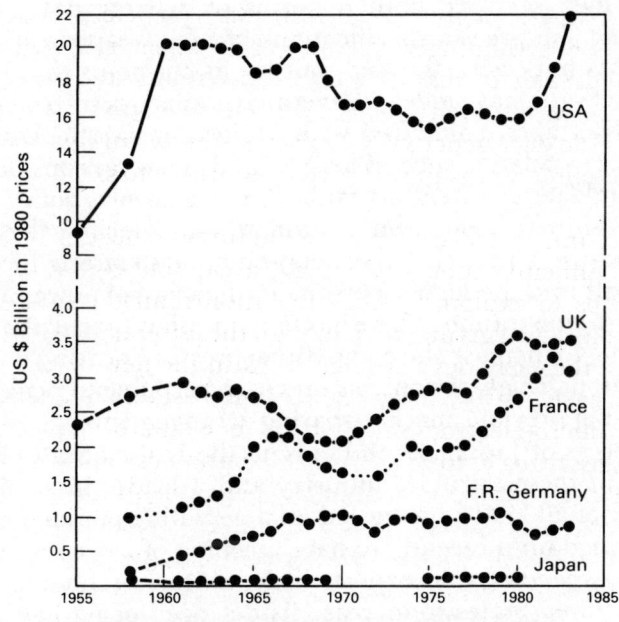

Source: *Armament and Disarmament Information Unit.*
Vol. 8, no. 3, May-June 1986.

FIGURE 2. Military Research and Development Expenditure: International Comparison

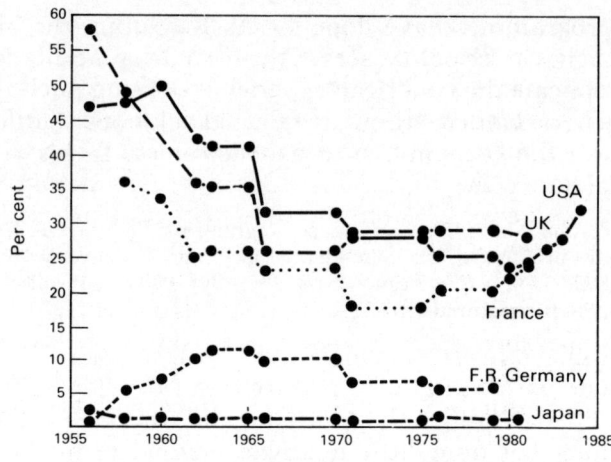

Source: *Armament and Disarmament Information Unit.*
Vol. 8, no. 3, May-June 1986.

FIGURE 3. Military R & D Expenditure as a Percentage of National R & D Expenditure

policy often dominated by arms sales considerations. On occasion, arms export considerations have taken priority over the needs of the French armed services, both in terms of delivery dates or even in determining the precise specifications of new weapons systems. The price paid is not easily assessed, but Ergas considers that France and the United States have at least generated a mixed record of advantages and disadvantages, compared with the less successful United Kingdom. France's relative success arises, in his view, in considerable part from general factors extending far beyond electronics policy. The great political legitimacy, operating autonomy and technical expertise of French end-users agencies, have combined with strong incentives for success built into the highly personalised nature of power and careers in public administration. There has been an effort, particularly over the past decade, to increase the competitive pressure bearing on suppliers, notably through tighter controls on costs, recourse to penalty clauses and by easing previous market-sharing arrangements.[32]

The success of Japan, second only to the United States in terms of output from the electronics industry and a leader in semiconductor technology of all kinds, has rested on a very different kind of strategy. Japan claimed until recently to have a ceiling of 1 per cent of GNP on all military spending and exports little military technology. Pressure from the United States to increase its defence budget has been interpreted by some Japanese commentators as an American device to burden their economy.[33] Japan's staggering success in the electronics field has been based on a partnership between government and industry, and the development of an integrated industrial policy with clearly defined objectives. Industrial strategy has been geared towards high shares of world markets, with associated high volume production. This has emphasised the diffusion of technological change—adapting and transferring technology from one industry to another and importing technology from other countries—within a consistently supportive government framework. After the War the government introduced incentives to productivity and quality, quality control methods imported from the United States being first used in the electrical and the telecommunications industry. In the 1950s and early 1960s, government assistance included tax benefits, assistance in investment and protection of markets. In the mid 1960s, the emphasis shifted to government sponsorship of industrial projects in national priority areas. In 1966 a programme for high performance computers was started and a functional element programme has been outlined for the 1980s. The Ministry of International Trade and Industry (MITI) invests heavily in civilian R & D, such as for the memory chip market during the 1970s.

It was the United States, investing heavily in military and industrial electronics, which provided the early push in the digital revolution and in integrated circuits. The Japanese, concentrating on consumer electronics, where costs were of prime importance, were not pioneers: to have invested early in research in integrated circuits for consumer goods would have been a massive act of faith. But none has been more adept than Japan at diffusing new technology into consumer electronics, in which it enjoys a commanding position.

Japanese electronics firms have been among the foremost of what the British Electronics Economic Development Council (EDC) described as, 'overseas competitor firms [which] have increased their share of the world market—often at British expense—by the application of consistent long-term business strategies which develop internationally competitive strengths'.[34] Britain, by contrast, has had a less clearly consistent strategy for electronics development, than either the Americans, the Japanese or the French, and major British firms are smaller not only than their major Japanese and American competitors but also than their European competitors. (GEC and Plessey ranked 14 and 21, respectively in the world in terms of group sales; see Figure 1.)

Britain had a healthy civil industrial base at the time of the Second World War, which enabled it to respond well to the challenges of radar, precision navigation, radio fuses and other military innovations. After the war, the R & D organisations established for the wartime drive in military technology, far from being dismantled, were extended. The close links between British electronics companies and Government defence departments were retained, and British peacetime R & D in electronics became characteristically concentrated on defence and telecommunications projects financed by military procurement contracts. British military funding directed research and development towards scientific investigations such as the properties of gallium arsenide, still only a specialist product, rather than process-oriented technological questions such as the development of design and manufacturing techniques. This emphasis was misplaced, even in military laboratories, and helps to support the observation that, 'The UK had probably developed more original technology than any other country besides the United States but it has very little to show for it.'[35]

British research and development, in contrast to that of the United States, was concentrated in the postwar period on the same organisations that had been active during the War, or on their direct successors. The British microelectronics industry did not enjoy the kind of volatile commercial environment that the Americans had. Although British technology retained a leading position in some technological areas, mainly military, it lagged seriously behind that of the United States, and by the 1960s British companies were completely absent in many of the major areas of microelectronics. Neither was there any major change in the structure of the industry such as had occurred in America: the big electrical and electronics companies, heavily concentrated on defence and telecommunications, were still dominant. British civil electronics suffered badly from Japanese competition: by the late 1970s the erosion of the British home commercial base was so serious that even the British television industry had virtually disappeared. Consumer electronics goods were supplied primarily from Japan, the United States, Europe, or even the new producers in the Pacific and elsewhere.

The British electronics sector has steadily declined in international competitiveness overall during the past two decades. A trade surplus of £106 million in 1963 was replaced by a deficit of £876 million in 1982.[36] The Electronics EDC commented in 1982 that:

> The United Kingdom's electronic industry has a number of technological and product strengths, but it is relatively weak in most of the businesses where world markets are growing fastest and amongst which are the best opportunities for growth in the future. This is reflected in growing trade deficits, in a declining share of world markets and in deteriorating company performance as measured by the real growth of firms which is very slow in comparison with firms in the U.K.'s major international competitor countries.[37]

Since that report was written, the international competitiveness of the British electronics industry hs deteriorated further. Between 1980 and 1984, the trade deficit increased over tenfold from £181 million to £2,052 million, with import penetration growing from 37 per cent to 59 per cent.[38] The deficit is largely accounted for by the performance of two of the fastest growing sectors, consumer electronics and information technology; that is, British manufacturers have not concentrated on leading edge business.

The main strength of the industry is the capital goods sector, an area dominated by military electronic equipment, where it was a net exporter. The defence electronics market has, at least until the present level funding of the defence budget, been a buoyant one for British firms. The Ministry of Defence takes directly 20 per cent of the output of the British electronics industry. In addition it takes 50 per cent of the output of the aerospace industry, producing missiles, satellites and aircraft, which also have a substantial electronics component.[39] In 1984–85 four British electronics companies—GEC, Plessey, Racal Electronics and Thorn EMI—were each paid more than £100 million by the Ministry of Defence, as were British Aerospace (Aircraft) and British Aerospace (Dynamics).[40] Overall defence spending in Britain, at $22.7 billion in 1985, was higher than that of France, West Germany or Italy (see Figure 4), and sales of defence equipment have been a steadily increasing proportion of total industrial output. The British defence electronics market is heavily protected: 80 per cent of the national defence electronics budget was spent with British firms. Dependency on the military market was particularly high in the radio, electronics and capital equipment sector.

The export market for British defence electronics, unlike that for civil electronics, has been growing. The Secretary of State for Defence, opening the British Army Equipment Exhibition at Aldershot in June 1986, claimed that Britain's share of the world arms market had increased from 7 per cent in 1984 to 9 per cent in 1985.[41] About 20 per cent by value of these exports are likely to be electronic subsystems: over the period 1979–83 this would represent an annual overseas sale of defence electronics of some $0.2 billion. But the percentage of electronics in defence exports is likely to be rising. Although military exports usually contain rather less electronics than the systems currently being delivered to the exporter's own armed services (because of military restrictions), the traditional reluctance to export high technology arms is believed to be weakening as the world arms trade becomes more competitive.

Research and development is of crucial importance to the electronics industry, linked as the industry is to continuous innovation. Britain spends more than France, Germany or Japan, all major electronics competitors, on military research and development, both in total and as a percentage of all national R & D expenditure. (See Figure 3.) In 1985 overall defence expenditures for major NATO countries were a massive $267 billion for the United States, $22.7 billion for the United Kingdom, $20.1 billion for France and $9.2 billion for Italy—

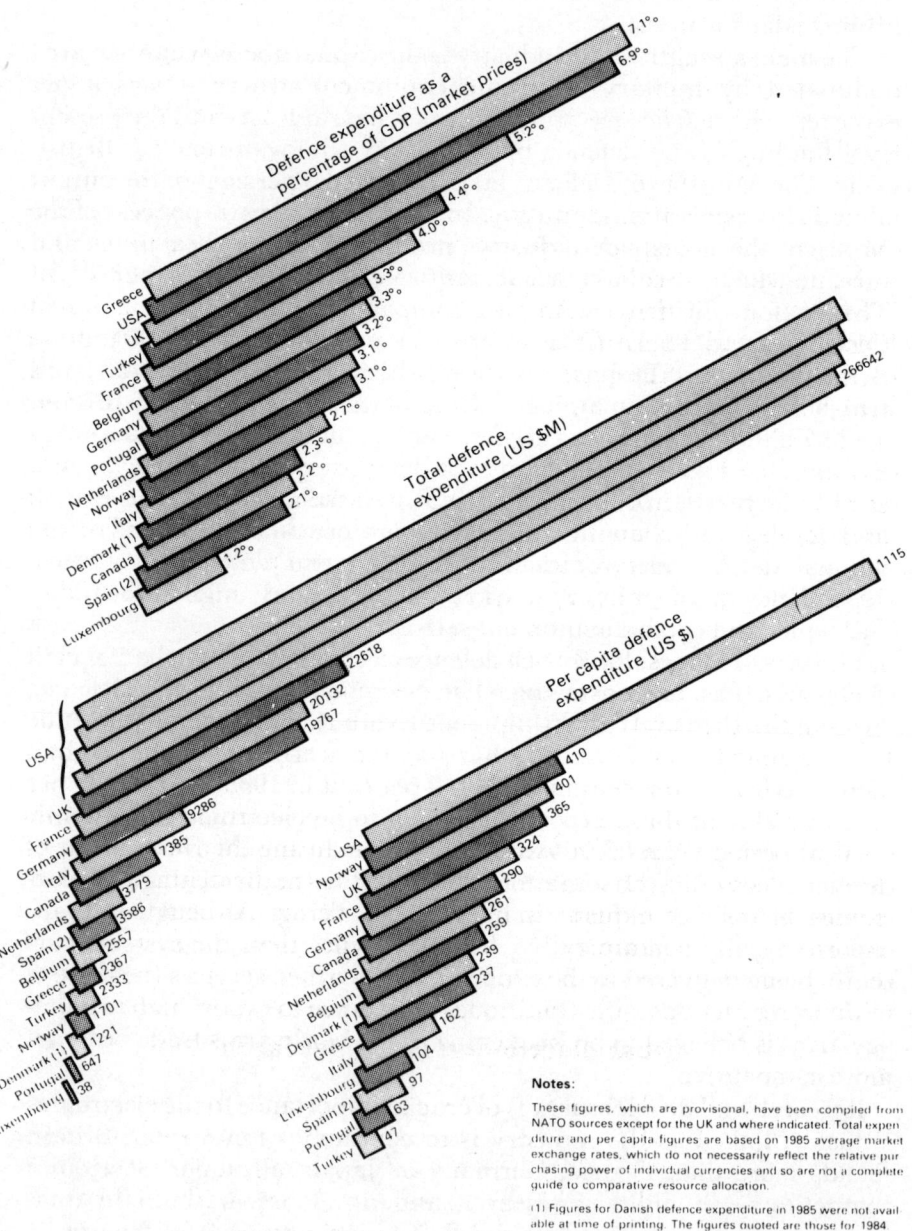

FIGURE 4. A Comparison of Defence Expenditure: NATO countries 1985

see Figure 4). A recent report by the Council for Science and Society explained the scale of Britain's military R & D in terms of the country's imperial history:

> As a former Great Power, and former leader in the development and production of weapons, Britain has inherited a large and sophisticated capability in military R & D, and a tradition of largely independent and self-sufficient development and production of the weapons needed to support her global role.[42]

The proportion of military R & D as a percentage of all R & D, already high in Britain by international standards, has been rising. The Ministry of Defence share of total Government funded R & D expenditure, under 50 per cent in the late 1970s, was 51.8 per cent in 1985–86.[43] This compares with a world average proportion of 25 per cent and an OECD average of 33 per cent in the late 1970s. In some parts of the British electronics industry, notably radio, radar and capital goods, two-thirds of all R & D was military. On the other hand, although electronics is the most R & D-intensive of all the industrial sectors in Britain, both British manufacturing industry as a whole and the electronics sector in particular invest less of their own funds in research and development than their major competitors. In British industry as a whole, less was spent on industrial R & D in real terms in 1983 than in 1978.[44]

The long-term characteristics of the British electronics industry, in international terms, have been its growing trade deficit; its concentration on the capital goods (especially defence) sector and on its protected home market; the relatively high proportion of military R & D; and the comparatively poor exploitation in the export market of its strength in defence electronics,[63] compared with France and the United States. The relationship between military and civil electronics has been the central concern of most research conducted during the 1980s. But, at a higher level, others have claimed that what has really distinguished the British electronics industry in international terms has been its lack of a coherent national strategy.[45]

British electronics: policy research

The poor showing of the British electronics industry and the distinctive relationship between civil and military electronics in the United Kigdom have not been thoroughly researched. There is, for instance, no extensive community of independent analysts of military R & D in Britain, as there is in the United States. There is a basic lack of empirical data and analysis. However, a number of different working parties have published reports, based on varying amounts of empirical research, during the last five years.

The Electronics EDC, made up primarily of industrialists with representatives of the trade unions and the Department of Industry, issued a major report in 1982. Its main recommendations stemmed from its conclusion that British electronics was disadvantaged, relative to its major foreign competitors, by the lack of an overall national strategy, formulated within the framework of a coherent, national, planning process. Its main recommendations were that:

> 1. Industry (including trade unions) and Government should together select the main actual and potential areas of business strength in the U.K. electronics industry on which a thrust into international markets can be built.
> 2. They should also work together to define strategies for each of the selected businesses so that the companies and public sector policies can be applied in a consistent and coherent way to build up the international competitive position of the UK firms.[46]

Among a host of detailed recommendations, the EDC report argued for a joint Ministry of Defence, Department of Industry, manufacturing industry and university programme, for the development of advanced computer systems.

The report argued that the atypical character of the United Kingdom electronics industry, with its heavy concentration on defence, had discouraged internal competitiveness.

> The procurement policies of British Telecommunications and the Ministry of Defence (which between them account for 40 per cent of the UK market) have not provided sufficient incentive for companies to concentrate their product range and thus gain manufacturing efficiencies. Whilst this has led to a broad base of expertise, it has diverted companies from developing business strategies based on world markets and fostering manufacturing activities on a world scale. The retreat, for example, from commercial and consumer markets for IT [information technology] products and the increasing dependency of firms on the in general more slowly growing capital sector markets is in part clearly accounted for by the type of relationship UK firms have with the public sector, their major customer.[47]

The EDC saw a need to harmonise procurement with broader industrial objectives and, in particular, to obtain greater technological 'spin-off' from defence to civil electronics. The EDC took the concept of spin-off at its face value, although it is, in fact, problematic. The usual notion of spin-off is that some new discovery is made in the course of research and development for a military project. The possibility of a commercial application becomes apparent and a product eventually results. Examples of this simple kind are, however, quite rare, and closer examination of what are claimed to be examples reveal a complex situation, involving various kinds of interaction between military and commercial sectors, rather than a straightforward one way flow. The EDC, however, apparently unaware of the complexities of the concept, recommended a study to identify barriers to the transfer of technology from military to civil areas and subsequently commissioned a study from Sir Ieuan Maddock, former Chief Scientist at the Department of Industry.

Maddock's brief was a restricted one: to examine the civil exploitation of defence technology in Britain. He identified four main types of company. Type A companies were research–development–design–production companies which have been concerned almost wholly with defence equipment for many decades (initially the War years followed by subsequent expansion). Their whole stance is governed by the requirement of the single, dominant customer—the Ministry of Defence—and much of their own decision making is subsidiary to that of the MoD. In the type B companies, the civil and the defence work are done side by side in the same design offices, laboratories and workshops. The only examples of type B companies identified were the specialist software houses that could move with ease between defence and civil work. Type C companies are those which have developed very advanced and unique technologies for applications primarily in the civil field and have then undertaken, very successfully, the application of some of this technology in the defence area. Type D companies operate almost wholly in the civil field, developing and marketing a range of distinctive products, some of which are subsequently sold to the MoD. They felt strongly that the armed forces should make much greater use of standard, civil products.[48]

Maddock was not optimistic about the general possibilities of military/civil transfer:

> It has to be faced that the likelihood of type A companies making a major contribution in the civil area (other than aerospace) is vanishingly small, and even strong measures by the government are unlikely to have more than marginal effect. Type B companies are more likely to generate civil business but . . . there are very few such companies (or divisions of companies) and their market perception is largely beamed onto large, high complexity, capital projects. Type C companies are far more likely to make effective commercial use of any new technological skills they develop or which are made available to them since the bulk of the enterprise lies in the civil area. Type D companies by definition are aimed at civil markets, but because they exist outside the main defence orbit, they are likely to have only limited access to the frontier technologies or have minimal opportunity to cultivate a new technique which would be of benefit in the civil field but also provide an important defence resource.[49]

Maddock argued that the organisational structure of the defence bureaucracy and of contractors inhibited technology transfer. An orientation towards military contracts influenced the structures of industries and firms, changing their organisational ability to adjust and compete in civilian markets. In particular, there was, in his view, a culture gap within firms which internally inhibited spin-off from military to civil work:

> The large multi-product British companies should themselves be able to make

> lateral technology transfers on the base of their own activities, but most admitted when pressed that this did not happen to anything like the extent that was desirable. Some argued that this was due to the highly autonomous management of the individual subsidiary companies, some argued that they were 'too busy' with their own tasks to look around elsewhere in the group and all admitted to a measure of 'not invented here'. . . . There can be little doubt that mobility of personnel is one of the most powerful methods of achieving technology transfer but there was very little evidence in the large companies that there was a *deliberate* policy of moving people between the defence and the civil fields. Indeed in most cases there was hostility to the idea because the people in the two fields were regarded as being 'so different' even to the extent of one individual stating emphatically that he would never use an engineer from the civil field on his defence work.[50]

Others have indeed argued that there are more shared values, standards and techniques between the R & D community, whether based in industry or government, than between the military and the civil division of a single company.[51]

Maddock's recommendations were aimed primarily at making defence technology available to his C- and D-type companies. As in the report of the House of Lords Select Committee on Science and Technology on engineering research and development of the same year, he argued that the benefits of military contracts were primarily confined to only a small circle of large firms, who kept knowledge gained from the resulting R & D largely to themselves.[52] His suggestions for breaking out of this pattern included the establishing of technology brokers to facilitate information exchange and technology transfer; the reintroducing of industrial applications units at government laboratories; the drawing up of directories of funding sources and technology development staff; the launching of an awareness campaign aimed at non-electronics companies which might benefit from new technology; and the creating of local liaison groups between defence laboratories and local industry. He argued that the MoD should be encouraged to use commercially available products where possible, and that there should be less vertical integration within the larger companies and more use of small, specialist sub-contractors. But Maddock was not optimistic that these measures would arrest the decline of the British electronics industry:

> The purpose of this study was to examine whether by some changes in practice and attitudes the undoubtedly high defence technology can be used as a device for reversing the decline of the civil electronics industry—an industry which is becoming central to all engineering and service industries. Whilst there are certain actions that can be taken . . . it would be naive to promise that these would have anything more than a marginal effect.[53]

Although there has subsequently been little specific study of electronics, an increasing number of researchers with a more general industrial focus have gone further than Maddock, not just to doubt the possibility of significant spin-off, but to argue that the dominance of the military sector damages the civil sector. Critics of military R & D levels in Britain emphasise the opportunity costs involved in the drain of scientific resources away from civilian uses. They deny that spin-off (from military to civilian applications) is an important phenomenon. They claim that military innovation focuses on new military products rather than on new manufacturing processes and therefore cannot be of much value to the civilian economy; and they argue that any gains that do accrue to the civilian economy could have been obtained more cheaply by means of projects dedicated specifically towards civil objectives in the first place.[54]

A study of the impact of defence procurement on the electronics sector in London prepared by the Greater London Trades Union Resource Unit,[55] for example, came to the conclusion that Britain's failure to reverse the decline in electronics and to diffuse and commercially exploit electronic technology stemmed from the sector's high level of defence dependency, which was the outcome of successive British governments' commitment to high defence expenditure:

> There is an urgent need for a reduction in defence spending and a planned transfer of the saved resources into key civil technology areas to stimulate innovation, diffusion and employment generation.[56]

Most of the other relevant research has been less specific to the electronics sector. Some two years after Maddock, a general report extending beyond electronics, entitled *The Economics of Defence Spending*, written by M. S. Levitt for the National Institute for Economic and Social Research,[57] claimed that there was an inverse correlation between military sales and productivity. Overall productivity gains for the years 1976–81 in the British electronics industry averaged 3.6 per cent a year, but gains in the radio, radar and electronics capital goods sector, which is especially tied to military sales, averaged −0.6 per cent a year over the same period. In contrast, the components sector, with only 6 per cent of its sales to military markets, averaged productivity improvements of 8.7 per cent annually. Levitt concluded that the investment, productivity and foreign trade performance of the defence-oriented part of the electronics industry compared with the parent sector revealed relatively low investment, relatively poor productivity growth, a relatively closed home market and a relatively poor export performance.[58]

The Council for Science and Society was, in its report, *UK Military R & D*, like Maddock, not optimistic about spin-off from military to civil areas: much military R & D was, in its view, inherently unsuitable for civilian application:

> Most military R & D leads to *product* innovation. Much of the innovation on which civilian industry depends is, however, in improvement in the manufacturing *process*

not in new product development. It is through process innovation that companies compete over price and quality.⁵⁹

Pointing out that the electronics industry, the most R & D-intensive of all British industrial sectors, had over half of its R & D funded by the Government, and overwhelmingly by the MoD, the report stated that:

> Looked at another way, UK private industry now spends less of its own money on electronics R & D than French industry or any other of its major competitors. British firms have failed to become leaders in most fields of semiconductors or other electronic developments over the past 30 years, with the major exceptions of military electronics and areas within electronic computers and electronic instruments.⁶⁰

The report concluded that Britain's military R & D, despite being geared to military roles determined by government, was excessive in relation to the country's economic status. It argued that R & D resources should be concentrated, but not necessarily by a reduced commitment to defence as such nor the abandonment of a major military role; more likely options were greater international collaboration, more dependence on imports, and manufacture under licence or co-production. It held out little hope that major economies could be achieved through a policy of increased competition as such.

A more radical report from the Science Policy Research Unit at Sussex University focused attention on the competition for skilled and scarce electronic engineers between military and the civil sector:

> The military sector ... appears to attract the cream of the technically skilled labour force since it provides work in exciting and prestigious fields. The drain on skilled manpower, particularly in electronics, has been a major factor inhibiting the expansion of high technology industries and, equally important, the take-up and use of new techniques in older industries.⁶¹

Noting the importance of the defence sector to the electronics industry, and particularly to electronics capital goods, the authors argued that:

> part of the reason for Britain's poor performance in the information technology sectors derives from these defence commitments. At a time when Britain should have been concentrating resources on meeting the challenges and opportunities presented by these new technologies, which involved both developing new industries and updating the design and performance of equipment in older industries, we have seen an increasing proportion of our manufacturing resources and particularly our highly trained manpower, pulled into the defence sectors.⁶²

This report concluded that Britain should reconsider whether she needed a national industrial defence base. It, in effect, argued that a major shift in national policy, involving a diversion of funds from defence to other sectors, was necessary to arrest economic decline:

> If Britain is to break the vicious circle of decline, an important precondition must be a reduction in the relative size of the defence sector and level of military R & D.

> ... Britain has got to re-orientate her technological effort away from the defence sector and towards the civilian market. However unpalatable politically, this task will not be achieved without a relative shift of resources out of the military sectors — in other words, substantial and lasting cuts in expenditure on defence equipment and R & D.[63]

However, Ergas's report *Does Technology Policy Matter?*, for the Centre for European Policy Studies, did not conclude that Britain spent too much on military R & D but that it managed it badly. Britain, like France and the United States, was what he called 'mission oriented', that is, primarily concerned with major projects of national importance, often with an emphasis on national defence. But the United Kingdom, like France, was less willing than the USA to disseminate the results of defence-oriented innovation.

> Once programmes are set up and running there is little external political pressure in favour of disseminating results; second, the members of the programme 'club' themselves have little interest in seeing results publicised and they tend to count more heavily in dissemination decisions and third, the external environment — and notably that in the universities — has been perceived as probably hostile and possibly untrustworthy. As a result, the information generated by mission-oriented programmes has tended to remain confined to a small circle of participants.[64]

Overall, the United Kingdom was less successful than either France or the United States in managing mission-oriented strategies:

> Mission-oriented research has tended to yield few direct benefits while possibly crowding out a substantial share of commercial R & D. The indirect spin-offs have been low, creating a 'sheltered workshop' type of economy; a small number of more or less directly subsidised high technology firms, heavily dependent on and oriented to public procurement; and a traditional sector which draws little benefit from the high overall level of expenditure on R & D.[65]

There was, in Ergas's view, too little competitiveness in the public sector market:

> The UK's major difficulties arise from the pervasive lack of incentives in its system of mission-oriented R & D. The propensity of British agencies to form 'clubs' with their suppliers — within which each supplier is treated on the basis of administrative equity rather than commercial efficiency — weakens whatever incentives suppliers may have to seek an early lead, while also ensuring that the resources available are so thinly spread as to be ineffectual. Finally, the reluctance to build penalty clauses into development contracts and to terminate unsuccessful contracts . . . aggravates an inherent tendency to cost over-runs.[66]

Successive British Governments do not have well-established procedures for the review of and response to reports originating from outside the civil service. There is, in particular, no body analogous to the Office of Technology Assessment in the United States. Science, technology and industry tend not to be at the centre of the government agenda: reports on such topics as the electronics industry are consequently picked up only at the margins. As we shall see in the next chapter,

Parliament, too, has neither the resources nor, in most cases, the specialist interest, to play a major part in pressing for a serious response to outside policy documents. Consequently, British policy towards civil and military electronics relates only marginally to the research summarised above, where its conclusions happen to coincide with the prevailing government ideology.

British electronics: policy implementation

A number of policy conclusions emerge from the research surveyed above. As we have seen, a number of analysts have diagnosed the exceptional dominance of British defence R & D expenditure as a key problem. It is clear, as the section on military electronics showed, that there is little prospect of reducing the R & D levels required to support modern, increasingly electronically-sophisticated weapons systems. Although a few radical commentators, notably Mary Kaldor, have argued for a move towards simpler, cheaper weapons,[67] all the signs, internationally, are that weapons will continue to become more costly and electronically sophisticated, and the rate of obsolescence will continue to increase, as the United States in particular invests vast sums in defence electronics.

There are two possible strategies for cutting military R & D in order to liberate material and manpower resources from military to civil electronics. One would be to reduce defence commitments and hence abandon one of Britain's more R & D-intensive defence roles (beyond the reductions already planned during the prospective period of level funding). This would necessitate both a reappraisal of defence objectives and a plan to deal with the resulting job losses, i.e. defence conversion. The second option would be to alter procurement policy by reducing national self-reliance in defence equipment, either by increased international collaboration, manufacture under licence or simply purchase from overseas. The reduction in military R & D that may be expected from international collaboration is limited. In practice, experience of collaborative projects suggests that political difficulties and conflicting requirements have produced inefficient, compromise solutions. There have been large time and cost penalties from the duplication of activities in several countries and it has proved difficult to cancel unsuitable systems. Kaldor and her co-authors point out that, 'the need to satisfy the Services of two or more different countries, and the insistence on maintaining broad national capabilities rather than agreeing on a true division of labour, have tended to encourage embellishment in design and inefficiency in production, while resulting in equipment that is not ideal for a particular role.'[68]

Alternatively, some analysts, and specifically Ergas, have focused on the lack of competition. In their view the calibre and the adaptability of management in the electronics sector have been sapped by their long-standing custom of making big profits in protected markets, such as defence and telecommunications, where orders have been doled out on the basis of 'Buggins' turn', rather than fought for in fierce competition.[69] According to this thinking a policy of increased competitiveness in defence and telecommunications might not only get better value for money in those markets, but promote the competitive fitness of the entire electronics sector and counteract the tendency of leading-edge companies to fall back on the soft option of defence contracts.

Although the Electronics EDC emphasised the need for a national strategy more explicitly than the other reports surveyed here, this is implicit in the recommendations of virtually all of them. On the other hand, virtually no-one was optimistic about the possibilities of greatly increasing spin-off from military to civil electronics.

How do these conclusions compare with the policies actually pursued by the British Government? First of all, government ideology has not been receptive to the kind of interventionist or even directive government implied by the strategic plan recommended by the Electronics EDC. Rather, its philosophy has inclined it towards a combination of policies for increasing competitiveness, encouraging arms exports,[70] sponsoring some joint civil/military research in advanced computing as recommended by the Electronics EDC,[71] and promoting spin-off from military to civil applications, in spite of the doubts about spin-off widely voiced in Maddock's and other reports. In addition there was some impetus, particularly during Michael Heseltine's term as Secretary of State for Defence, towards more collaborative defence projects.

The Government's encouragement of British firms to participate in research for the American SDI was an example of this strategy; a special office set up within the Ministry of Defence, including officials from both the Department of Trade and Industry and the Department of Education and Science, was given the task of identifying and promoting relevant British skills and technological capabilities available to industry, universities and research institutions. Participation would, the Government claimed, 'enhance the United Kingdom's ability to sustain an effective British research capability in areas of high technology relevant to both defence and civil programmes.'[72]

Given the doubts frequently expressed by researchers about spin-off from military to civil developments, it is not surprising that the claim that civil benefits would flow from the SDI programme was disputed. Robert Jackson, Conservative M.P. for Wantage, argued in the *Daily Telegraph* that:

The Government has estimated that by 1988 Britain will be short of some 5,000 graduates in the vital information technologies—yet, with the emerging defence technologies, not to mention SDI, it seems clear that the lion's share of these essential people will be absorbed by defence industries where they will yield relatively poor economic value.[73]

John Chowcat of ASTMS went further: he argued that the civilian spin-off from SDI research was likely to prove small-scale at best and that technological and economic interdependence on the USA would create a new impetus for 'brain-draining'. In this context, the vulnerability of electronics would be further exposed.[74]

In the event, British firms have won few of the contracts for the SDI. In 1987 these amounted only to 45 million dollars, compared with the £1 billion which Michael Heseltine, then Secretary of State for Defence, held out hopes of achieving.[75] What is more, direct attempts to exploit the British Government's support for SDI in order to secure other electronics export orders to the United States have not been conspicuously successful. Mrs Thatcher's direct appeal to President Reagan, in a letter of September 1985, asking that British backing for SDI should be borne in mind before the American Government awarded its major contract for a battlefield information system, did not prevent the contracts being awarded to the French Rita, rather than the British Ptarmigan system.[76]

The Government has conspicuously not attempted to reduce military R & D, in spite of some support for this from Conservative backbenchers. The Government has, instead, emphasised its commitment to spin-off: as the 1984 defence estimates put it:

> We are concerned ... that technology generated in [defence] industry in this way at the taxpayers' expense should be available for exploitation more widely in non-defence fields.[77]

Of the various ways in which spin-off might be encouraged, the Government has favoured private, or a mixture of private and public initiatives. In October 1985 a private venture, Defence Technology Enterprises (DTE), launched with venture capital provided by Lazard merchant bank and other City institutions, was given unique and privileged access to four military research establishments: the Royal Signals and Radar Establishment at Malvern (an internationally renowned centre for defence electronics), the Royal Aircraft Establishment at Farnborough, the Admiralty Research Establishment at Portland and Portadown, and the Royal Armament Research and Development Establishment (RARDE) at Fort Halstead, Sevenoaks. The aim was to identify promising areas in defence research that offer lucrative commercial spin-off. Initially 400 cases of technologies with commercial potential were fed into the database of the RARDE for dissemination to some 140 client companies at home and overseas.[78]

Since it was set up in 1985, DTE has acted as a broker 'in about 20 negotiations for manufacturing licences'.[79] For example, the Malvern Program Analysis Suite (MALPAS), developed at the Royal Signals and Radar Establishment, has been licensed to Rex Thompson and Partners. MALPAS checks for a variety of types of error in software. At a recent Alvey Club meeting on formal approaches to software,[80] it was stated that MALPAS was available only for the language CORAL 66. CORAL 66 is predominantly used by British defence contractors, and so the release of MALPAS as product is of major benefit to defence work — hardly a good example in relation to Maddock's aims. 'Boom' is available for licensing but this too has predominantly defence applications as it simulates the effect of weapons such as armour-piercing warheads. A software product available for licensing that does have civil applications is 'Scribel' which is a word processor developed in British defence laboratories. These three examples of software illustrate that products from defence laboratories may go to support defence work in the companies rather than to support or supplement products in other industrial sectors. Products that are of use in civil applications may force direct and fierce competition from civil development and the military derivate is not assured of commercial success. Thus far there are few examples of 'negotiating licences' leading to major civilian products.

Experience so far, therefore, leaves it open to question whether an emphasis on the civil exploitation of defence electronics is necessarily the most cost-effective approach. Initiatives such as the formation of DTE may be a useful way of finding and exploiting the results of military R & D but it has yet to prove cheaper than putting comparable resources directly into civil R & D. Searching out ideas arising from military R & D and seeking a commercial user may involve higher costs and greater uncertainties than directly funding civil research and development to meet a clearly specified goal.

A central plank of Government policy towards electronics, as to other sectors of industry, has been the promotion of more vigorous competition. In the civil sector the privatisation of British Telecom was followed by the opening of the hitherto protected British telecommunications market to international competition: British Telecom decided to pit AXE digital exchanges developed by Sweden's L. M. Ericsson against System X, developed by Plessey and GEC. In the military sector, Heseltine oversaw an initiative designed to produce greater competition in defence procurement in 1983. The procedures for military tendering were made more competitive. The number of cost plus contracts — described by Admiral Sir Lindsay Bryson as 'a licence for technical and scientific orgies'[81] — were reduced. This major reform, whose importance should not be underestimated, has put increased pressure on contractors to hold down costs and is said to have already produced very substantial savings in the defence budget, for instance, in the field of remote ground sensors.[82] The Government claims, more speculatively, that it will also have substantial benefits for the international competitiveness of British industry.[83]

There is, however, little sign as yet of any real opening up of the procurement market to smaller subcontractors, as recommended in the Maddock report, and it has become evident that a policy of increased competition necessarily conflicts to some extent with the continuing desire to sustain a domestic defence industry on strategic and economic grounds. It is not clear how far it is possible to increase competition without opening up military procurement to international tendering and abandoning the belief that national capabilities must be maintained to develop and produce all types of equipment for security of supply.

Heseltine has spelt out the importance of the subsidy involved in maintaining a national defence capability:

> there is practically nothing you cannot buy cheaper from the United States of America because they have huge production runs, huge resources, huge research programmes, funded by the taxpayer, and if we want to cut down Britain's industrial capability all we have to do is to go to the United States of America and they will enable us to buy products cheaper, and they are very good products...[84]

So total commitment to competition seems out of the question. Even attempts to increase competition somewhat, in the broad context of a national defence capability, are fraught with dilemmas. Negotiations over supply contracts can be long and complicated, and it is necessary for the Ministry of Defence to work closely with contractors in the early stages of research. If the number of potential suppliers is increased, in the name of competition, this raises the costs of the contracting process and prolongs it. Each potential supplier incurs a cost in negotiating the contracts and the aggregate cost of all the suppliers involved in the negotiation obviously increases as the number of potential suppliers increases. Ultimately, with a single purchaser, as in the case of the MoD, it is the purchaser who bears the costs of the suppliers, so that the cost of both successful and unsuccessful transactions for each possible supplier will be passed on to the purchaser in this or later contracts. Transaction costs will therefore rise with an increased scale of competition. Dealing with fewer suppliers not only reduces transaction costs, it improves the volume of sales for the vendor and may bring about some economies of scale.

As far as the electronics industry is concerned, it was the Nimrod case above all which brought home the difficulties of promoting competition in the context of a national defence-procurement capability. The project was so large and the technology so specialised that it was necessary to form the GEC/British Aerospace consortium to undertake the project in the first place, and, when it got into difficulties, only largely American consortia had the capability of offering satisfactory alternatives. Similarly, Marconi Underwater Systems Ltd., a subsidiary of GEC, is currently in a monopoly position for torpedo development and production in the United Kingdom.

Attempts to increase competition inevitably cut across a multitude of strategies, including a certain cosy clubbiness, which have grown up over the years to cement relationships between existing contractors and civil servants involved with defence. These strategies had gone some way to create an integrated supply and user industry, which may in effect operate to maximise profits overall.[85] One such strategy is the transfer of personnel, usually from the Ministry to the defence industries, but occasionally, as in the spectacular instance of Peter Levene, from the defence industries to the Ministry. Such practices greatly complicate the question of how to increase competition.

The policy of increased competition in weapons procurement may cut across the efficient management of a total weapons system. A defence equipment-procurement programme may have an initial competitive tendering stage, but after the award of the contract it becomes progressively more expensive to find an alternative supplier. The chosen supplier then has no incentive but to reduce prices and, if the price is increased, the government has little room for negotiation. The Ministry has tried splitting contracts into several stages, for example, separating the development contract from the production contract. However, these approaches increase the number of transactions, and hence the cost of completing the project, create complicated interfaces, and place the purchaser in the possibly inappropriate role of overall project director. Noting that the MoD had not found a solution to the problem of managing the total weapons system, the report of the Comptroller and Auditor General on the torpedo programme suggested that the possible creation of a single, prime contractor to manage and control the total weapons system could make a major contribution to overcoming the associated problems, but that it ran the risk of further reducing competition.[86]

The problems and tensions inherent in a policy of enhanced competition have also been apparent on the civil electronics front. Britain's pioneering role in opening up its telecommunications market to wider international competition has placed its industries in a more exposed position, because none of its neighbours has yet reciprocated.[87] The Government's policy of promoting more vigorous competition within the United Kingdom came at a time when the imperatives of survival in world electronics markets seemed to require economies of scale in production beyond the present scope of even the largest British electronics companies. The tension between the desire to maintain competition, particularly among the major defence contractors, and the pressures towards creating larger units of production, was crystallised in the proposed merger between GEC and Plessey, a merger supported by the Department of Trade and Industry and opposed by the Ministry of Defence.

Plessey and GEC, both major defence contractors, accounted between them for about two-thirds of British defence electronics, and, in one category, air-defence radar, controlled almost 80 per cent of national production.[88] The Ministry of Defence formally opposed GEC's takeover bid on the grounds that it would reduce competition for defence contracts. Defence Ministry officials feared that, if GEC were to acquire Plessey, the same kinds of problem that had dogged the torpedo programme and the Nimrod programme, the latter delayed because of problems with the GEC radar, would spread to Plessey. This company was responsible, for example, for the Ptarmigan communications network and development of underwater sonar for the Trident submarines.

But the logic of a GEC–Plessey merger had been unexpectedly floated a year earlier by Lord Lucas, a junior trade minister, in the course of a House of Lords debate. Officials of the Department of Trade and Industry, whose job it is to promote the electronics and the telecommunications industry, gave evidence in favour of the merger to the enquiry of the Monopolies and Mergers Commission. They argued that both companies were small by international standards (GEC, the larger of the two, was, for instance, less than a quarter of the size of General Electric in the United States), and that the merger was necessary if the United Kingdom was to retain a domestic electronics capability to compete in the international market place dominated by the United States and Japan. The cost of developing and marketing big telecommunications and defence systems was, they argued, becoming so high that only companies which could finance long-term development and maintain high volumes of output were likely to remain profitable. In the DTI's view, the MoD could maintain competitiveness by being prepared to buy from abroad. British Telecom, for its part, supported the merger on the grounds that bringing together the Plessey/GEC System X businesses would make for lower costs, greater efficiency and increased export opportunities. The Mergers and Monopolies Commission opposed the merger on the grounds that the benefit of it to the System X business would not outweigh the adverse effects in other market areas and the loss of competition for defence electronic equipment.[89]

The proposed GEC–Plessey merger was only one of a series of episodes, of which the best known is probably the Westland affair,[90] in which civil and military interests, as interpreted by the Department of Trade and Industry and of Defence, respectively, collided. The tensions between military and civil interests were illustrated again in 1986 when the MoD's response to cost overruns in the torpedo progamme was to persuade three or four more British companies to compete for a business with virtually no export market prospects and no civilian applications. Who could have foreseen, asked the *Financial Times*, that 'the MoD would go absolutely barmy in the name of competition.'?[91] Clearly, the manifold problems attached to reconciling a policy of increased competition in the defence-procurement field with other defence policies and with other aspects of government policy will not be easily resolved.

The continued implementation of the policy of increased competition itself requires governments to maintain, over the long term, a resolute resistance to the lobbying of defence industry interests. If such a policy is to be maintained, it necessarily involves squeezing some defence contractors. Signs that this is now the case were born out in recent report on the effect of government policies on the electronics industry, conducted for the City firm of Wood Mackenzie:

> The value for money policies being applied vigorously by the Defence Ministry are likely to create difficulties for some contractors, leading to takeovers and mergers in the electronics sector. . . . Changes in the previously cosy relationship between the MoD and its defence contractors are turning the defence electronics industry into an area of high risk and high reward.[92]

Similarly, a recent survey organised by the Defence Industry Quality Assurance Panel argued that there were increased delays in awarding contracts as a result of the MoD's attempts to increase competition between contractors. Recent results announced that the British software group, Systems Designers, showed reduced profits, which the group blamed in part on delays in the awarding of MoD contracts.[93]

If competition is to be effective, the Government has to be prepared to allow less competitive companies to go to the wall. The problem is that these firms will naturally lobby for more regulation to protect them from failure. This is one of the few occasions when management and trade unions can be relied on to unite, forming a lobby of formidable political influence. If there is simultaneously a continuing assumption that a broad defence base is essential for national security, then the chances are that eventually these companies will get their way.

Conclusion

Little detailed research has been carried out on the electronics industry as a whole. However, some trends seem beyond dispute. Electronics will become an increasingly important, increasingly complex and increasingly costly component of modern weapons systems. The range of choices which this opens up in military terms will immensely complicate the management of defence, where large-scale investment in military R & D, especially in electronics and information technology, is creating a more complex, more unpredictable, and possibly more uncontrollable military environment. It is by no means clear that electronic aids to decision making are keeping pace with the increasing complexity of the tasks involved, both in defence procurement and in military planning. This chapter was concerned primarily with the economic rather than the military problems caused by the increasing sophistication of military electronics. It seems clear that neither Britain nor France, nor indeed the United States, yet shows signs of facing the major policy, strategy and procurement dilemmas which flow inescapably from the current heavy investment in research and development in military electronics.

The difficulties for the British in particular will be exacerbated by the problems of the electronics sector as a whole. Military and civil interests are treated politically, bureaucratically and in budget terms as separate issues and as such are often seen to pull in conflicting directions. The MoD has a great deal of power and exerts much influence over the British defence industries, but it receives no budgetary incentives of any kind to take broader economic questions into account when making major procurement decisions. Civil and military interests cannot, of course, really be separated in this way. A healthy electronics base is essential for national security, conceived in its broadest sense, as some analysts have explicitly recognised.

The research which has been done is by no means exhaustive or definitive. But it is clear that there is little match between research conclusions and the policies actually implemented. The Government has, typically of governments elsewhere, confined itself to adopting those, usually small, parts of research proposals which fit in with its political ideology. The reform of procurement procedures, the introduction of greater competition in the protected British defence and telecommunications markets, attempts to increase spin-off and the initiation of some joint Government–industry–university technical research were, indeed, all recommended in studies of the electronics industry. But they were only part of a larger set of proposals. Reports have emphasised the importance of formulating an overall national strategy on civil–military relations and electronics R & D, yet these, perhaps more fundamental but ideologically less appealing proposals, have been largely ignored by the Government. Policies of increased competition may, on their own, produce unintended consequences and may exacerbate the apparently conflicting interests of the military and the civil sector. The policy of competition can best be seen as an experiment: seen in this light, it is essential that the effects of the policy are continuously monitored and that the results of information is made available, so that full and independent evaluation may take place.

The policy of spin-off, itself a complicated concept, rests on some conception of shared civil and military interests. But while spin-off may provide a way of offsetting some military costs, it is never likely to stave off the problems posed for the defence budget by increasingly sophisticated military electronics. Nor is it the starting point for a revival of the British electronics industry. For that, something more than pragmatism is required.

Notes

1. Fred Guterl, 'Today's navies: plying a sea of debate', *IEEE Spectrum*, vol. 19, no. 10, October 1982, p. 48.
2. While it is true that more sophisticated, advanced generations of a given type of tank, ship or aircraft cost more per unit than the less sophisticated, older types, it does not necessarily follow that the relative unit costs of standard equipment rise, and there is some American evidence that they do not. See M. S. Levitt, *The Economics of Defence Spending*, National Institute for Economic and Social Research, Discussion paper no. 92, May 1985, p. 35.
3. *Statement on the Defence Estimates 1979*, Cmnd 7474, HMSO, 1979, p. 17 and *Statement on the Defence Estimates, 1987*, Cm 101–II, HMSO, 1987, p. 52.
4. 'World military expenditure, the past decade', *SIPRI Yearbook 1981*, Taylor & Francis, 1981, p. 7.
5. *Statement on the Defence Estimates 1985*, Cmnd 9430-II. HMSO, 1985, p. 17.
6. Ibid.
7. 'Defense spending for electronics still strong'. *Defense Electronics*, November 1985.
8. *Defense Electronics*, vol. 18, no. 7, July 1986.
9. Adrian Milne, 'The Malvern link', *The Listener*, 10 May 1984, p. 12.
10. 'Department of Defense software initiative', Communications of the ACM, August 1986.
11. Frank Barnaby, *The Automated Battlefield*, Sidgwick & Jackson, London, 1986, pp. 162–4.
12. Norman R. Augustine, 'Brilliant missiles on the horizon', *IEEE Spectrum*, vol. 19, no. 10, October 1982, pp. 96–7.
13. Quoted in Gunter Friedrichs and Adam Schaff, *Micro-electronics and Society: For Better or Worse*, Pergamon, 1982, pp. 254–5.
14. Barnaby, see note 11, p. 51.
15. In Britain, the Phoenix remotely-piloted vehicle is being developed for introduction in the early 1990s. *Statement on the Defence Estimates 1986*, Cmnd 9763-I, HMSO, 1986, p. 32.
16. Robert Bernhard, 'Electronic countermeasures', *IEEE Spectrum*, vol. 19, no. 10, October 1982, p. 59.
17. Larry W. Sumney, 'VHSIC: a promise of leverage', *IEEE Spectrum*, vol. 19, no. 10, October 1982, p. 94.
18. Quarndon Stock List, September 1986, Quarndon, Slack Lane, Derby.
19. *Washington Post*, 12 September 1984.
20. Richard D. DeLauer, 'The force multiplier', *IEEE Spectrum*, vol. 19, no. 10, October 1982, p. 36.

21. A technology such as SOS, which has limited commercial potential, may not sell in sufficient volume to justify commercially the establishing of a second source, to ensure continuity of supply of components. If the primary investor has military interests then the market potential of the product may be limited by export restrictions and the possibility of licensing the process may be prohibited. The military involvement limits the potential commercial market and may exclude some possible second source deals across national boundaries. The restricted possibilities for second sourcing make it less likely that commercial buyers will use spin-off from military technology.
22. *Jane's Fighting Ships 1981–82*, Jane's Publishing Company, 1981, p. 129.
23. Mary Kaldor, *The Baroque Arsenal*, Deutsch, 1982.
24. Barnaby, see note 11, p. 3.
25. Richard Perle, 'Analysis', BBC Radio 4, broadcast on 4 February 1987.
26. Jolyon Howorth, 'Consensus of silence: the French Socialist Party and defence policy under Francois Mitterand', *International Affairs*, vol. 60, no. 4, Autumn 1984.
27. Quoted in: L. Soete and G. Dosi, *Technology and Employment in the Electronics Industry*, Francis Pinter, 1983, p. 164.
28. Quoted in: R. W. de Grasse, *Military Expansion, Economic Decline: the Impact of Military Spending on U.S. Economic Performance*, Armouk, M. E. Sharpe Inc, 1983.
29. R. Smith and G. Georgiou, 'Assessing the effect of military expenditure on OECD economies: a survey'. *Arms Control*, vol. 4, no. 1, May 1983.
30. Henry Ergas, 'Does technology policy matter?', Conference on the Economics of Technology Policy, Centre for European Policy Studies, 1–2 September, 1986. A revised version will be published as 'The importance of technology policy' in *Economic Policy and Technological Performance*.
31. R. Gilpin, *France in the Age of the Scientific State*, Princeton University Press, 1968, p. 292.
32. Ergas, see note 30, p. 9.
33. R. P. Smith, 'The significance of defence expenditure in US and UK national economies', *Built Environment*, vol. 11, no. 3, 1985, p. 164.
34. Electronics Economic Development Council, *Policy for the U.K. Electronics Industry*, National Economic Development Office, 1982.
35. E. Braun and S. Macdonald, *Revolution in Miniature: the History and Impact of Semiconductor Electronics*, Cambridge University Press, 2nd edn. 1982.
36. Philip Gummett, 'Controlling military R & D'. *ADIU Report*, vol. 8, n. 3, May–June 1986.
37. Electronics EDC, see note 34, p. iv.
38. Mary Kaldor, Margaret Sharp and William Walker, 'Industrial competitiveness and Britain's defence', *Lloyds' Bank Review*, October 1986.
39. *Statement on the Defence Estimates 1986*, see note 15, p. 46.
40. Ibid.
41. *Financial Times*, 24 June 1986.
42. Council for Science and Society, *U.K. Military R & D*, Oxford University Press, 1986, p. 28.
43. Cabinet Office, Annual Review of Government Funded Research and Development, 1987, HMSO, table 1.1, p. 7.
44. John Chowcat, 'The SDI business', *ADIU report*, vol. 7, no. 5, September–October 1985.
45. Electronics EDC see note 34.
46. Ibid.
47. Ibid.
48. Sir Ieuan Maddock, *Civil Exploitation of Defence Technology*, Report to the Electronics EDC, 1983.

49. Ibid., pp. 10–11.
50. Ibid., pp. 7–8.
51. Council for Science and Society, see note 42, p. 48.
52. House of Lords Select Committee on Science and Technology, *Engineering Research and Development*, HL Paper 89, 1982–83, paras. 19.1 and 19.2.
53. Maddock, see note 48, pp. 7–8.
54. Judith Reppy, 'Military R & D and the civilian economy', *Bulletin of the Atomic Scientists*, vol. 41, no. 9, 1985, pp. 10–4.
55. *Lost Jobs Wasted Skills: the Impact of Defence Procurement on the Electronic Sector in London.* Prepared by the Greater London Trades Union Resource Unit, for Greater London Conversion Council, 1986.
56. Ibid., p. 28.
57. Levitt, see note 2.
58. Ibid.
59. Council for Science and Society, see note 42, p. 46.
60. Ibid.
61. Kaldor *et al.*, see note 38, pp. 45–6.
62. Ibid., p. 48.
63. Ibid.
64. Ergas, see note 30.
65. Ibid.
66. Ibid.
67. Kaldor *et al.*, see note 38.
68. Ibid., pp. 43–4.
69. Ergas, see note 30.
70. *Statement on the Defence Estimates, 1986*, see note 15, p. 48: 'We have always encouraged companies to put forward proposals that will both meet our requirements and have good export potential; but we are now planning to include in defence contracts specific provisions to cover export possibilities.'
71. For example, the Alvey Programme, started in 1983, a £350 million programme for industry, the MoD and academic groups to pool research expertise in areas of advanced computing.
72. *Statement on the Defence Estimates 1986*, see note 15, p. 5.
73. *Daily Telegraph*, 3 July 1986.
74. Chowcat, see note 44, p. 13.
75. Hansard, vol. 124 no. 62, column 149, 8 December 1987.
76. *Sunday Times*, 10 November 1985.
77. *Statement on the Defence Estimates, 1984*, Cmnd 9227-I, HMSO, 1984, p. 20.
78. *Daily Telegraph*, 3 July 1986.
79. At the time of writing, Defence Technology Enterprises had acted as a broker 'in about 20 negotiations for manufacturing licenses'. *Financial Times*, 11 June 1987, p. 10.
80. Peterborough, 10 March 1987.
81. *Daily Telegraph*, 29 November 1985.
82. *Statement on the Defence Estimates 1986*, see note 15, p. 47.
83. *Statement on the Defence Estimates 1985*, Cmnd 9430-I, HMSO, 1985, p. 35.
84. Quoted in: Lawrence Freedman, 'The Case of Westland and the bias to Europe', *International Affairs*, vol. 63, no. 1 1987, p. 6.
85. Cf. comment on GEC's purchase of a government shipyard: 'A group so beholden to MoD favour can do itself no harm by helping out with what could be a troublesome divestiture programme.'
86. *Ministry of Defence: The Torpedo Programme*, Report by the Comptroller and Auditor General, 21 March 1985. HC Paper 291, 1984–85.

87. *Financial Times*, 27 June 1985.
88. Guy de Jonquières, 'Everything to play for', *Financial Times*, 21 July 1986.
89. *The General Electric Company PLC and the Plessey Company PLC*. A Report on the proposed merger by the Monopolies and Mergers Commission, August 1986, Cmnd 9867 HMSO.
90. The Westland affair also demonstrated that a policy of increased competition could conflict with the parallel policy of increasing collaborative defence procurement projects with Europe. Freedman, see note 84.
91. *Financial Times*, 29 September 1986.
92. *Defence Electronics: Fighting Back*, Wood Mackenzie & Co., May 1986.
93. *Computing*, 30 October 1986.

6

British Defence Decision Making: the Boundaries of Influence

MARGARET BLUNDEN

Introduction

Mr Malcolm Muggeridge, when asked in a television interview in Britain in September 1973 what was the most boring topic for an article he could imagine, said 'After NATO what?'.[1] The tedium which Muggeridge experienced when defence questions were forced upon his attention typified prevailing public attitudes in Britain during most of the 40 years after the Second World War. Except for a brief period in the late 1950s and early 1960s, defence appeared to be of only specialist concern. The inter-party consensus on defence meant that elections were determined very largely on domestic issues.

The limited public concern for defence questions had obvious advantages for defence decision makers, since major decisions could be taken by a small circle of decision makers, and managerial strategies for weapons procurement need not be buffeted by contentious political debate or overt industrial lobbying. During the late 1970s, however, the anti-nuclear challenge mounted to long-established defence policies, and the collapse of the inter-party defence consensus, combined to make defence—or at least nuclear weapons—the major public issue which it has since remained. Whereas the older defence interest groups such as Pugwash fed relatively little into any public defence debate, new organisations critical of nuclear weapons and/or the NATO alliance, founded or revamped in the late 1970s and 1980s, became, like other single-issue campaigns, steadily more professional at carrying out research, providing public information, and at regular lobbying and briefing which specifically targeted Members of Parliament.[2] Some upholders of the existing defence paradigm now felt compelled to attempt to influence public opinion on defence questions and to start similar organisations of their own.[3] The House of Commons, for its part, established a Defence Committee in 1979, and held in January 1980 its first debate on nuclear policy for 15 years.

In the United States where, like the United Kingdom, the environment of defence decision making has become more lively and questioning, the processes of decision making themselves are said to have become more participative. There, it is argued, 'the system by which society decides to research, develop, build and deploy its weapons is becoming far more open'.[4] How far is this also true of the United Kingdom? This chapter looks at the processes of British defence decision making—the management of civil–military relations at the highest level—in the second half of the 1980s, at the kinds of thinking and expertise which feed into them and at how far they are changing. It starts with a hypothesis about the composition of the defence decision making system. Whom should one take to be the decision makers?

Although it is politicians who set the tone for policy discussions at the working level, parliamentary influence over decision making has not traditionally been strong in Britain, nor indeed in most other Western European countries. A report compiled in 1979 for the Western European Union (WEU) argued that 'national parliaments and their defence committees, with the exception of Germany and the Netherlands, are usually inadequately informed on defence matters' and concluded that 'while party political pressures have over the years affected the size of national defence budgets, parliamentary pressures never seem to have clearly affected, either positively or negatively, any government's major procurement preferences or any decision to alter the shape and deployment of the armed forces'.[5] In Britain as elsewhere in Western Europe where, the WEU report argued, 'many expensive projects not properly controlled are hidden under research and development heads',[6] some major nuclear procurement decisions, large and significant enough to amount to policy making, have in the past been taken by secret cabinet committees, without reference not only to Parliament but to most of the senior members of the government responsible for them. The best known of these were the Attlee decision to begin the British nuclear force in the late 1940s and the Heath/Wilson/Callaghan programme to provide a new 'front end' for existing Polaris missiles (the Chevaline programme) of the 1970s.

After Parliament belatedly learned about the Chevaline programme, the Ministry of Defence (MoD) agreed in 1982 to give the House of Commons Public Accounts Committee confidential information about all projects on which it had made a commitment to spend at least £200 million (subsequently revised upwards to £250 million). However, an escape clause written into the agreement by the then Comptroller and Auditor-General allowed for cases in which a government department could ask the Comptroller not to pass on information to Parliament on the grounds that it could damage the national interest. In this way it was possible for the Government to conceal from Parliament the existence of the Zircon listening satellite programme, until recently being developed by British Aerospace and Marconi at an estimated cost of £500 million. In 1987, after the disclosure of the Zircon programme, the House of Commons Defence Committee set a new requirement, with the agreement of the MoD, for separate reporting to them on an annual basis of projects expected to cost more than £25 million for development or £50 million for production.

Despite these changes to financial procedures, the British Parliament still does not have anything like the detailed control of the defence budget enjoyed in the United States by Congress, where even top secret or 'black' programmes, such as the Advanced Technology Bomber (Stealth), are voted by Congress although they are kept secret from the public. The British defence estimates do not provide the kind of detailed costings for individual weapons systems which allows Congress to vet carefully what money is to be spent on particular systems before it is allocated. Michael Hobkirk, former British civil servant, reports that defence officials in the United States have been heard to speak with wonder and envy of the ease with which their British colleagues can get budgets approved by the legislature.[7] But no national parliament in any other NATO country has anything like the power of the United States Congress to decide the respective size of the three armed services, and to distribute funds for the use, maintenance, development and procurement of their military equipment.

The pressure of the House of Commons Defence Committee inside Parliament, and of the public debate generated as defence has become a major issue outside Parliament, have squeezed more information out of defence ministers. The Government appears more sensitive than hitherto to public commentary. Mrs Thatcher's Government has not been subject to the specific political pressures for secrecy on nuclear matters bearing on previous Labour Governments, embarrassed by internal divisions on nuclear policy. However, it remains constitutionally the case that the British Government prepares and decides; there is no legal requirement to consult and Parliament's role is only advisory. In practice, although increased amounts of information have been made available, especially through the Open Government documents, information is usually given after decisions have been taken in order to justify them (as in the case of Trident), rather than before the event to allow wider and more informed thinking to contribute to the decision making process.

So this chapter assumes that it is still ministers, civil servants in the MoD, the Foreign and Commonwealth Office and the Cabinet Office, and senior military officers who constitute the defence decision making system in Britain; and that Parliament, along with the defence industries, defence analysts, and single-issue organisations in the defence field can best be described as pressure groups in the environment. British defence policy normally reflects the thinking of only a small number of people. What attitudes, capacity and experience now frame that thinking? How, if at all, are the processes of decision making changing? How open are they now to outside comment and in what way are they influenced by expert or scientific opinion?

The decision makers

Ministers

Defence, as Anthony Eden, the former Prime Minister, remarked and as Margaret Thatcher still exemplifies, is very much a prime minister's special subject.[8] Indeed, the post of Minister of Defence was created in 1940 by the then Prime Minister, Winston Churchill, for himself. All prime ministers are inclined to be proprietorial about defence, although few of them since Churchill had previously held a defence portfolio.[9] This may not matter much, since the role of the prime minister is to take an overview of national policy. The contribution of the prime minister should be to integrate defence policy into a broad vision of the nature of the international system, Britain's role within it and the interconnections between internal and external affairs. Such a contribution was exemplified in France by President de Gaulle, who placed French defence policy within a grand design encompassing the strengthening of France's internal institutions, colonial disengagement and relations with the Third World, the creation of a European confederation and the bolstering of medium-sized powers *via-à-vis* the superpowers. Few British prime ministers, however, have articulated a view of defence in the context of such a broad overall policy, apart from the possibly mythological special relationship with the United States.

The Secretary of State for Defence, for his (no woman has yet held the office) part, does not hold the most prestigious of cabinet posts: ambitious ministers may see it as a staging post and will certainly have their own political agenda. Responsible as they are for a budget of some £19 billion, they are liable to be envied by the other spending ministers, and it is moreover sometimes difficult to relinquish the job without political discomfiture. Recent Secretaries of State, particularly John Nott and Michael Heseltine, may feel that things have not changed all that much since the days of the War Office, described in 1903 as 'a place where the most strenuous endeavours meet with the smallest return, and from which little is to be got save vituperation'.[10]

Defence ministers, with one or two exceptions such as Denis Healey and Fred Mulley (now Lord Mulley), both Labour, continue to have less specialist expertise themselves than their counterparts in many other Western countries. The holders of the defence portfolios (Secretary of State for Defence, two Ministers of State, and two Parliamentary Under Secretaries of State), are not necessarily or even usually drawn from the small number of defence specialists within Parliament. It may even be, as Max Hastings has said, that, despite the increasing public concern with defence, 'few defence ministers are sincerely interested in national defence',[11] and that well-publicised initiatives such as the 1983 reorganisation of the Ministry of Defence may have more to do with establishing a ministerial reputation than with meeting defence needs.

Ministers in the British system are in theory gifted amateurs. Constitutional practice normally rules out the possibility of recruiting to ministerial office experts from outside Parliament, as is common in the United States and elsewhere in Europe.[12] There is, for instance, no British parallel to the career of the Norwegian Johann Jorgen Holst, a former junior minister who became head of the Norwegian Institute of International Affairs and subsequently returned to public office as Minister of Defence. And a system like the French, where ministerial office is incompatible with membership of Parliament, produces ministers like Charles Hernu, the author of numerous defence publications[13] and André Giraud, a previous director of the *Commissariat à l'Énergie Atomique*, responsible for civil and military nuclear programmes.

In the British defence policy making system, it is civil servants and senior military officers, doing tours of duty at the Ministry of Defence, who are the experts. Senior civil servants, unlike ministers, are defence professionals, selected like other civil servants from among the most academically able graduates of their generation, although more likely than other civil servants to have a science or engineering degree.

How does this distribution of expertise affect the kind and quality of decision making? It certainly will not ease the minister's task of controlling his department. The modern secretary of state struggling, like his Victorian predecessors, to reconcile competing commitments with limited resources, may still recognise what Florence Nightingale, speaking of the then War Office, described as 'a very slow office, an enormously expensive office, a not very efficient office, and one in which the minister's intentions can be entirely negatived by all his subdepartments, and those of each of the sub-departments by every other'.[14] But it is sometimes claimed that the British system, with its amateur tradition, has its advantages. John Nott, Conservative Secretary of State for Defence from 1981 to 1983, argues that, surrounded as the secretary of state is by immensely knowledgeable civil servants, it took him only about three months to understand the issues and that the advantages of going in 'cold' are an open mind, a willingness to listen and a capacity to adjudicate between conflicting counsels.[15]

This is a managerial view of the role of the defence minister. It assumes that his job is to best manage the existing system, or at least to try, as Nott himself did, to accomplish change within the existing parameters, rather than to try to change the parameters themselves. Exceptional ministerial skills would be required if an incoming government were committed to changing the parameters of defence policy, as established since the Second World War. But in any case, the task of decision making for major weapons procurement programmes is a demanding one.

First, it involves making assumptions about the international political and military environment for some 10 to 15 years ahead. Apart from the think tanks within the MoD itself—the Concepts Division, the Directorate for Defence Policy and the Secretariate (Policy Studies), the minister will be advised by the Foreign and Commonwealth Office, whose planning staff is specially charged with long-term thinking and which formed the Arms Control and Disarmament Research Unit in 1965 to stimulate long-term perspectives. Secondly, defence decision making has major domestic implications for industry and technology policy, and is linked with such variables as the strength of the merchant marine, and home affairs such as civil defence and terrorism. Weapons systems and the research and development programmes which yield them are inextricably involved in the competition for priorities within public spending as a whole.

The difficulties of reconciling these two aspects of defence decision making were memorably articulated, in the context of the United States, by Samuel Huntingdon more than 20 years ago:

> The most distinctive, the most fascinating, and the most troublesome aspect of military policy is its Janus-like quality. Indeed, military policy not only faces in two directions, it exists in two worlds. One is international politics, the world of the balance of power, wars and alliances, the subtle and the brutal uses of force and diplomacy to influence the behavior of other states. The principal currency of this world is actual or potential military strength: battalions, weapons and warships. The other world is domestic politics, the world of interest groups, political parties, social classes, with their conflicting interests and goals. The currency here is the resources of society: men, money, material. Any major decision in military policy influences and is influenced by both worlds. A decision made in terms of one currency is always payable in the other.[16]

Apart from its political complexities, defence is also a daunting managerial task, which involves harnessing industry to marry up uncertain and constantly developing civil/military technologies with what are often vague and fluid military missions. Aspirations to anything approaching efficiency in this area demand an understanding of processes, of how to manage change within complex organisations which are either resistant to change or liable to manage change in their own way and sometimes according to their own objectives. In particular, ministers need to understand the need to plan for following through changes, which will otherwise drain away into the sands. This is no easy task, given their normally brief duration at the MoD.

Then there is the day-to-day running of the armed services. The armed forces and their supporting civilian establishments are by many criteria the largest and most complex organisations in the country.[17] The Ministry has to channel instructions from the elected government to the armed forces and the requirements of the armed forces to the government; to turn the government's decisions and policies into operational plans and orders for the forces; and to organise the personnel, logistic and procurements requirements of the three services.[18] In 1985 the Ministry had a military and civilian payroll of over 550,000, spent a budget of £17 billion and was British industry's largest single customer.[19] But few ministers have much managerial experience and they are handicapped by their often short periods in office compared with the long time-scales of defence procurement. As Sir Frank Cooper, Permanent Under Secretary of State for Defence from 1975 to 1983, sees it from a civil service perspective, the idea that ministers can actually manage this complex system is a myth:

> We have had 22 ministers of defence since 1945 and the casual way in which you have six ministers of state for defence procurement in seven years, is beyond disgrace. It is quite ludicrous to say that in a democracy you change people who are supposed to be in charge, in a ministerial sense, of a budget of eight plus billion pounds. So you are very light on knowledgeable defence ministers. They try very hard, they nearly all in my experience work very hard, they listen, they try to come to a sensible view. But defence is a professional business and I think our system of ministerial appointments now is a powerful reason why we have been in decline, in that ministers have not stayed in jobs very long, they tend to come from broad-brush backgrounds and they have had no experience of management. The idea that ministers can manage anything cannot possibly be treated seriously because none of them have got any management experience, except by exception.[20]

Mismatches of time-scale bedevil the management of weapons procurement. First, at the level of political decisions, there is the difficulty of matching the long time-scales of weapons procurement and the subsequent operational life with the much shorter time-scales within which political and foreign policy predictions can confidently be made. Decisions about major weapons programmes, intended to be operational into the next century, fit ill with the political preoccupations of individual governments, concerned above all with electoral credibility within a five-year maximum time span. Second, at the management level there is the mismatch between the usual duration of ministerial appointments and that of major weapons programmes. Former civil servant Clive Ponting[21] sees this mismatch as a major obstacle to effective ministerial control:

In terms of defence procurement I think it is difficult for any one minister to get a grip on the programme. It is an enormous programme; they only see the very top of it coming up. If they are here for two years they will probably see a programme once while they are there and they will be asked to take one tiny decision on it, whether it is, say, to go into project development or to authorise production on the completion of development or whatever, and their room for manoeuvre on that is pretty limited. It is very difficult to see where ministers can make any real impact in controlling the defence equipment programme, how far anyone does. It is one of those sort of juggernaut-like machines that tend to run on. The ability of any one person to control it is minimal.[22]

Civil servants

This is not to say that civil servants can effectively control procurement either.[23] They too are generalists who move within different areas of the MoD—or even other parts of Whitehall—according to their planned career structure, and may suffer similarly from lack of continuity. But they are powerful at all levels of both procurement and policy making. To start with, they greatly outnumber ministers: in 1986 there were 203 senior civil servants of assistant secretary grade or above, compared with only five political appointments (a number quite insufficient, in American terms, to control a modern defence department, in spite of the fact that the government of the day controls civil service promotions to the highest level).

Senior civil servants have access to classified information—sometimes better access than junior ministers. They understand how the large and unusual Ministry of Defence conglomerate works[24] and it is second nature to them to be policy relevant. They do at least have some more continuity than either ministers or their military colleagues, who may only be doing two- or three-year tours of duty. They have unrivalled—some would say jealously guarded—links with other government departments. The grip which senior civil servants have over policy making is thought to have been strengthened by the 1983 reorganisation of the Ministry of Defence, designed to streamline the defence policy making structure by integrating the top military command and top civil servants in a combined defence staff, although some argue that the reorganisation has actually increased the military input into policy making, at the expense of civilian. In any event, it has had the effect of limiting the number of people with real power in the policy making process.[25]

The influence which defence civil servants have over ministers derives from the assumption—often a myth—that their advice derives from an expertise which is impartial, whereas the advice of the military is tainted with self-interest. Nott's reflections are illuminating in this respect:

> The civil service has much more power than the military, because they were much more experienced in strategic terms. The military came and went. A man came in as minister for two or three years, and did the job and then went out again. Whereas there was a great body of knowledge on the part of the civil servants who were all defence specialists and, in terms of strategic understanding and an objectivity of resource allocation between the services, they were much superior to the serving officers. The serving officers were very good, they were all very intelligent and they argued their case very well but basically they were arguing their own corner. But I found the civil servants on the whole the key to the answers, because they had been thinking about these things all their lives and were trained to think in an analytical way, whereas on the whole the services are not trained in quite that way.[26]

Although civil servants may appear both more analytical and more objective than the military, their mindset is still largely structured by their shared social background and education, by their much stronger dialogue with English-speaking defence professionals abroad than with fellow Europeans and others, and by the norms and expectations of their own organisation. Existing defence policies will tend to seem commonsense to them, unless they run into managerial or financial problems. The civil service, like military or other organisations elsewhere, are socialised into particular ways of seeing and doing. Defence civil servants, as the main source of continuity, are moulded by the traditions of British defence policy making, in the sense defined by G. M. Dillon:[27]

> A tradition in policy making is an interpretation of experience, more or less codified and ritualised in a recognisable rhetoric and in established practices. It is lodged in a collective memory and accessible to future generations of decision makers to whom it is transmitted by a variety of devices.

The tradition which shapes civil servants on the policy making side of the MoD and in the Foreign and Commonwealth Office may usefully be seen as an interlinked set of strategies, norms and assumptions.[28] The guiding norm, substantially unchanged for a generation, is that Britain ought to retain as much of her former international influence and independence as possible; a central strategy, paradoxically, that the best way to sustain such influence and independence has been through dependence on the United States and the fostering of the 'special relationship', pre-eminently expressed in nuclear and intelligence matters. There has apparently been an assumption that Britain and the United States—more than, for instance, Britain and the rest of Western Europe—share a common commitment to certain core values and common interests, although this assumption has begun to be questioned recently.

The force of tradition, as encoded in its strategies, norms and assumptions, is strong in the Ministry of Defence which, like most European bureaucracies, has traditionally had little career mobility in either inward or outward directions[29] and, consequently, little challenge to received ways of thinking. Informal communication is also limited by the Official Secrets Act, which restricts contacts between policymakers and the outside world, and in procurement matters, by the demands of commercial secrecy.

Part of the changing defence environment of the 1980s has indeed been the willingness of a few retired senior civil servants, notably Sir Frank Cooper, Lord Zuckerman and Sir Ronald Mason, to contribute some of their privileged knowledge to the public defence debate. Cooper's lecture to the London School of Economics in 1986 even called for more open government, a Freedom of Information Act, and more mobility in and out of the civil service.[30] Such senior figures command public attention and enjoy a certain licence since it would be highly embarrassing for any government to attempt to prosecute them under the Official Secrets Act. Civil servants still in post have indeed been increasingly seconded to academe or well-known think-tanks in Britain or North America, such as the International Institute for Strategic Studies based in London, or Harvard University and the RAND Corporation in the United States. Secondments *out* to elsewhere in Europe or *in* the Civil Service are still rare.

In this respect things have changed only little since 1962 when Professor Laurence Martin, surveying the British defence policy making system, concluded that, 'There is undoubtedly a good deal of prejudice in Britain against anything which seems to be in breach of the principle of keeping affairs within the official family' and drew attention to 'the liabilities of a closed system which obscured the proper basis of debate, encourages rumour and constrains the exchange of ideas between the professionals and intelligent observers'.[31]

The British civil service, which recruits people in their twenties for what is expected to be a life-long career, has in the past resisted moves towards a more American style, where outside experts move freely in and out of Government, as administrations change. Denis Healey, one of the most experienced and knowledgeable defence ministers since the war, met with little success when he attempted to change the British system. His 1967 defence white paper announced the setting up of a Programme Evaluation Group, to include skilled outside advisers as well as civil servants and military leaders, but the Group was dissolved after only one year. Laurence Martin, then Professor of War Studies at King's College, London, has described the frustration of another of Healey's efforts to bring in a fresh perspective. After considerable discussion with Healey he received the formal offer of a post at the senior rank of Assistant Under-Secretary to establish a small, long-range planning staff to serve the Minister directly. While Martin was giving final thought to 'whether such a graft could take', the sudden general election of 1970 resulted in the return of the Tories. Within 24 hours a deputy secretary invited Martin to lunch to tell him not to accept the offer, although it probably could not have been withdrawn at that stage. The men were friends and the issue was not personal; it was just made very clear, if only by the speed with which the opportunity presented by the Minister's defeat was seized, that the civil servants had greatly disliked his initiative.[32]

Since that time, governments have managed to overcome a little of the traditional resistance of the civil service to outside recruitment. Mrs Thatcher's government set out to increase efficiency by bringing in executives from the defence industries on the procurement side. The appointment of Peter Levene, former Chairman of a major defence contractor, United Scientific Holdings, to head the Procurement Executive, was confirmed, in spite of its being technically in contravention of civil service rules. The expertise acquired by this sort of appointment is at best a much needed managerial experience and understanding of the defence industries — the acumen of the poacher turned gamekeeper. But it is not likely to provide what Frank Cooper calls the desperately needed 'intellectual background of explanation in order to convince people you've got to make changes'.[33] And, of course, the dilemma is that experience of management and knowledge of the defence industries is injected into the Ministry of Defence at the risk of encouraging some more questionable features of a bureaucratic/industrial complex.[34]

There has been little other new blood. In the mid 1980s there was some idea of appointing defence academics on short-term contracts, as the Ministry of Defence itself began to suffer from an outflow of its own personnel to the City. This idea was not unreservedly welcomed by academics, said by some civil servants to be fearful of losing their 'academic purity'.[35] There were also doubts about the level at which outside academics would be recruited, a crucial matter for the influence which their thinking could have in this hierarchical system.

The only senior position, apart from Head of Defence Export Services, usually filled by an outsider is the influential one of Chief Scientific Adviser[36] and this post was technically demoted during the Heseltine reorganisation of 1983. The Chief Scientific Adviser has Permanent Under-Secretary status, chairs three important committees — the Defence Research Committee, the Equipment Policy Committee and the Operational Analysis Review Committee — and may call on scientific staff at all headquarters and research establishments for information and advice. He (again, no woman has yet held the post) has direct access to the Secretary of State — indeed, his influence largely depends on the quality of that relationship.[37]

The scientist as adviser has been much less studied in Britain than in the United States, where he is to be found at practically every level of government. But it is clear that British Chief Scientific Advisers face difficulties stemming from the nature of the role. Although it is their scientific expertise which opens the door—civil servants being, it is said, more willing to concede their lack of scientific than of political or policy expertise—Chief Scientific Advisers rarely give advice on purely scientific matters. Like scientific advisers elsewhere, they are called upon to make judgements on a complex of scientific, technological, industrial and political factors, especially in relation to weapons research and development, which itself derives, in theory, from overall strategic considerations. The Chief Scientific Adviser is in effect 'a political adviser with an FRS'.[38] But scientific eminence does not necessarily confer political judgement. Chief Scientific Advisers may be inclined to overvalue the technical at the expense of other criteria. More fundamentally, they may fall for the myth of the 'value-free' expert, as crisply delineated by Harold Laski: 'the expert fails to see that every judgment he makes not purely factual in nature brings with it a scheme of values which has no special validity about it'.[39]

Then there are specific difficulties arising from the peculiarly isolated position of British Chief Scientific Advisers. Recruited as they are from outside the civil service, they do bring fresh thinking into the MoD. But once inside they quickly become as isolated as other civil servants from broader scientific thinking. This may be unavoidable to some extent since the scientific defence community is not strong in Britain. There is indeed the Defence Scientific Advisory Council, made up of academics and industrialists with some officials and chaired by an independent member; together with its subgroups this may total some 200 scientists.[40] But the generation of scientists who went into defence during the Second World War,[41] many of whom continued to take an informed and authoritative outside interest in defence after the war, has not been replaced. The loss of a pool of 'outside scientists' able to take on the Ministry of Defence on its own terms, seems particularly significant when one recalls how many of these former wartime scientists, such as Solly Zuckerman, P. M. S. Blackett, R. V. Jones and Joseph Rotblat, were critical of official thinking and able authoritatively to challenge the received wisdom. The American scientific defence community has not been depleted in this way. There the community has continually been replenished, keeping in being what has been called 'a sort of loyal opposition among the scientists'.[42]

As a result, British civil servants claim that there is little alternative to in-house analysis of procurement options which have an important technical dimension. This at least is the view of Sir Ronald Mason, Chief Scientific Adviser from 1977 to 1983 and formerly Professor of Chemistry at the University of Sussex:

The number of people I could call on for disinterested advice, even in a limited technical sense, outside the Ministry of Defence was very small. You would have to go through a very elaborate process of briefing them, just for the sake of involving them. Issues in defence technology, particularly strategic issues, belong to a very arcane corner. You would not actually expect very many people to be equipped to do it. It is true that in the United States you get people out of industry and national laboratories who are able to take on the specialists on their own terms. In the United States there is always somebody in a university or industry who has spent some time in government, who has spent some time in a government research laboratory. There are academics in the United States who can take on the Department of Defense—for instance, on SDI there is a scientific community who could argue with conviction that it is nonsense. That is not true here. I could not look to anyone who could tell me what an SDI architecture is, who could tell me, with real depth of understanding, of the feasibility of certain objectives.[43]

So the thinking of defence decision makers—ministers and senior civil servants—still takes place within a somewhat isolated and enclosed community. The closed nature of British defence decision making and the key role of senior civil servants within it, can be glimpsed in the case of the Polaris replacement decision, technically a procurement matter, but one which raised policy issues of exceptional importance. Here, the traditional inclination to deal with issues, especially nuclear ones, 'in-house' was reinforced by pressures for secrecy emanating from the politicians.

The background studies for the replacement of Polaris were conducted within the civil service, isolated from the broad debate of the issue than being conducted by policy institutes, defence academics, industry and pressure groups. As it happened, the early official and academic debates coincided. Late in 1977 Ian Smart, a former member of the British diplomatic service who was then Deputy Director of the Royal Institute for International Affairs, wrote for the Institute a paper which he intended to prepare the ground for a large-scale debate. He concluded that a nuclear-powered, missile-firing submarine (SSBN) fleet of five boats equipped with missiles having multiple independently-targetable re-entry vehicles (MIRVs) was the best solution.[44] This initiative coincided with a discreet suggestion by MoD civil servants to the Callaghan Government, that serious analysis of the questions should now begin. It is significant that it was Smart, one of the few members of the British academic defence community with civil service experience, who was able to anticipate both the timing of the issue and the criteria which would largely determine the choice of system. However, from the moment that the question began to be considered within Whitehall, all participation of civil servants in the public debate ceased. The civil servants who had asked to attend Ian Smart's seminar at Chatham House itself were withdrawn, on what is believed to have been a Cabinet-level decision. The issue was considered so sensitive, especially for a Labour Government, that it was thought essential to avoid signalling in any way that this was something the Government was interested in.

Inside Government, two study groups were set up to tackle what, as we have seen, is the immensely complex political and managerial task posed by major decisions of this kind. One group, chaired by Sir Anthony Duff, Deputy Under-Secretary at the Foreign and Commonwealth Office, considered the international consequences of renewing or discarding the British nuclear force. This corresponds to one side of the Janus-face of defence decision making referred to earlier, that is, the making of assumptions about the international political environment for some 10 to 15 years ahead. Simultaneously the other study group within the Ministry of Defence, presumably anticipating the conclusion of the Foreign Office group, considered which system should be chosen, that is, they had responsibility for the managerial or instrumental aspects of the decision. Both groups reported to a secret Cabinet committee consisting of the Prime Minister, the Chancellor of the Exchequer, the Foreign Secretary and the Secretary of State for Defence.

This division of labour did not reflect modern decision making theory, which recommends the continuous and interrelated consideration of ends and means.[45] It seems to have discouraged an iteration round the loop of which kind of new system, of what level of capability, and bought from whom, would have which kind of international and arms control consequences—an iteration particularly necessary since the Ministry of Defence group finally opted for a major potential increase in defence capability. Even more seriously, the briefs given to the two study groups, and their restriction largely to the foreign and defence departments, although the Cabinet Office and the Treasury were involved too, did not do justice to the other side of the Janus-face: 'the world of domestic politics, the world of interest groups, political parties, social classes, and their conflicting interests and goals'. What was involved were systemic changes extending well beyond the concerns of the Foreign Office and the MoD. The scale of the sums to be spent on a new strategic weapons system[46] had implications for the rest of the defence budget, for British industry and for the economy. The possibility of selecting a system with greatly enhanced capability had implications for domestic opinion, as well as international opinion and for arms control. The possibility of changing from Polaris to the enhanced counterforce capability of the Trident C4, could be thought, in spite of any official disavowals,[47] to have implications for British strategic doctrine.

However, neither group was asked to look specifically at the domestic implications of the decision, nor at the question of building a domestic consensus behind it, since this was the province of politicians, not civil servants. The possible permutations of choice to which the sums envisaged could be put, within the defence area, were not explored on an equal footing. The Chiefs of Staff were asked to rank their priorities amongst major defence missions and, thus boxed into a corner, felt constrained to put the maintenance of the independent nuclear force in first place.[48] The consequences of each of the possible strategic systems for employment and the British economy were not costed out to everyone's satisfaction. As David Greenwood, the University of Aberdeen defence economist and former MoD civil servant, commented to the House of Commons Defence Committee:

> that major procurement choices in the United Kingdom on the whole are not accompanied by that explicit display of likely employment and industrial consequences which one would like to see and which would enormously contribute to better informed decision-making and debate about decisions. I think that is a great pity.[49]

The structure of the decision making, its partitioning along department lines (excluding for instance, the Department of Trade and Industry), apparently reinforced the weight of tradition and the force of existing strategies, norms and assumptions within the Ministry of Defence group. The memory of the success, in procurement terms, of Polaris, based on what was then the most advanced American technology, and of the technical problems and cost escalation of the British Chevaline project, convinced the Ministry of Defence that the best way to serve the norm of British independence was by the strategy of dependence on the most advanced American technology. The secrecy of the studies reduced the complexity of the task, in that the scope for lobbying by British industrial interests was lessened.

The atomistic structure of decision making encouraged the Ministry of Defence group to see its brief in technical terms—as finding the most efficient means to an agreed end, within financial and technical constraints, including a desire for commonality with the United States. It strengthened the inclination to stress technical rather than other, less quantifiable criteria. The members of the MoD group concentrated their search for a solution, as one would expect,[50] in areas close to the old solution. They concluded quite early on that any successor to Polaris must be based at sea, and went on to opt for a sea-launched, ballistic missile system purchased, like Polaris, from the United States. The group considered, but did not rate as a serious difficulty, the fact that Trident provided a capability in excess of what was needed for minimum deterrence and would prove to be a controversial decision both nationally and internationally, particularly if a later decision were to be taken to change from the Trident C4 to the newer, more expensive and much more capable Trident D5. The argument that it had the capability to cope with the unforeseen throughout the 20 years of its life, was thought to be a more powerful argument. The 'unforeseen' was conceived of in technical rather than political terms. Major changes in the international arms control environment, emanating from superpower agreements, were not apparently anticipated.

The secrecy of the proceedings of which members of the House of Commons Defence Committee were subsequently to complain encouraged the neglect of the domestic dimension. In the view of Jonathan Alford, then Deputy Director of the International Institute for Strategic Studies and a former Army officer:

> There is a great danger that the cloistered atmosphere of Whitehall does inhibit free thinking, and tends to insulate you from the sorts of ideas which are going on around the place. There should have been more discussion about what people who could bring some expertise to this particular sort of problem would have told them: that they have got a problem if they go with Trident, and the problem is that people will say, it is an excess of capability over need. I do not think they recognised that because the technical information on which they made the decision all drove in one direction, the biggest bang for the buck was Trident, and they ignored some of the political factors to which they ought to have listened, which ought at least to have made them pay more attention to other choices.[51]

The military

The Trident case exemplifies the continuing key role of ministers and civil servants as the defence policy makers in Britain. Although the Chiefs of Staff were involved in the replacement study, their role was a constrained one. In practice, a number of educational and institutional factors limit the influence of the military over defence decision making.

Although some serving officers now see themselves as managers of peace rather than as captains of war, the emphasis, in all three armed services, is still on training for command, not for contributing to the analysis and formation of defence policy. Indeed, many outside analysts feel this is exactly as it should be. Command is what officers understand and to hold the appropriate command at the relevant age is the route to advancement. In the early part of their careers, young officers are required to concentrate heavily on that part of their profession that has to do with leading men and with managing fighting units, with their increasingly complex equipment. If they stay in the services and get promotion, the later stages of their careers will involve postings to the Ministry of Defence, an unfamiliar arena where the military sciences begin to overlap with the political arts. Here an experience of command is less useful than an understanding of political influence, if they want to do more than merely give technical or tactical advice. The transition from service in the field to a desk job in Whitehall is not easy to make, and some senior officers who would like to see the military exerting more influence on policy argue that it is not well managed.[52]

There have long been mechanisms for introducing military officers to the non-military aspects of national defence so important for policy making. The Royal College of Defence Studies, specifically amongst the military colleges, was set up in 1927 to educate British and Commonwealth students in strategic, political, economic, social, industrial and financial issues. The Ministry of Defence also operates a scheme whereby a small number of officers (11 in 1986) spend time at a selected university, taking courses or writing a thesis. A Belgian study of the European defence sector concluded in 1986 that 'the British military is more open-minded than most of its European counterparts'.[53] Across the board, however, British serving officers, especially the engineers, trained as they are to a high standard in the professional aspects of their work, do not have as broad a background in politically based studies, such as international relations, as their American counterparts. (This is partly because, no matter how distinguished they may become, they cannot hope for positions equivalent to that of American National Security Adviser, which would give them direct influence over policy making outside the formal structures of the Ministry of Defence.) Nor is the British military élite put through the intense academic preparation of French military high-fliers. Indeed, after a period of teaching at the Royal Military Academy, Sandhurst, Hew Strachan, Fellow of Corpus Christi College, Cambridge, concluded that the academic programme was superficial, that priority was given to the practical aspects of officer training, and that the spirit of open-minded enquiry essential to the learning process was suppressed. 'Lateral thinking is discouraged; consideration of NATOs demise or of territorial defence or of a European defence community is not evidence of originality but of heresy.' This meant that 'No more can we look on the British army as an institution for the advancement of learning'.[54]

Strachan's was explicitly a personal view, and reflected broader disagreements then current both about the way military history should be taught and about the most appropriate stage in the military career for academic enquiry. However, it does seem that unorthodox thinkers are made at least as uncomfortable in the services as in other large organisations. Not enough is done to encourage officers to participate in academic defence debate and to express their views by writing for journals. The military colleges do indeed undertake research—the Royal Naval College, for example, is well known for its work on maritime strategy, and the Soviet Studies Centre at Sandhurst is an important centre for the examination of Soviet strategy and tactics[55]— but it is difficult to assess their impact as much of their research is classified.

The inflexibility of the British service career limits opportunities for officers to take time out later in their careers for study and reflection on policy questions. One of the attempts to set up a research base at the Royal United Services Institute ran into difficulties, partly for this reason and partly because of funding and personality difficulties. As Farooq Hussein, former research director at RUSI, saw it:

> I did want to bring young officers, serving officers, in to do research work, which is done in the United States, where officers who are serving and have time between postings and so on spend six months, three months a year, or whatever in research institutes so that they become familiar with the sort of thinking that takes place around their force. If they become a chairman of the Joint Chiefs of Staff, they understand very closely because they have spent time at Harvard, they have been actively inserted into this community which forms American policy during the course of their career. And while everyone thought this was a good idea here, it actually is in practical terms impossible to take a serving officer in the British forces out of the system and put him here at RUSI to do some research work for a short period. There is no tradition of doing that here and if you want to change things like that you have, not only to be revolutionary-minded yourself, but to get everybody else to be revolutionary-minded at the same time, which is a rather hard thing to do.[56]

Impossible or not, there is a lack of opportunity for the military to step back from the pressures of service affiliation to take a broader look at national defence policy. This limits their contribution to major issues of defence policy. Admiral Sir James Eberle, former Commander-in-Chief, Naval Home Command and currently Director of the Royal Institute of International Affairs, agrees with Nott that ministers tend to rely on their civil servants rather than their military advisers. In his experience, ministers:

> mistrust the military, as not really being able to stand back and understand the broader context. And indeed a lot of the military are like that, they take the view that their job is just to give military advice and do not sufficiently appreciate that military advice has no substance unless it is taken within a political context. It is a message that I have been banging on about to my service colleagues for a long time, that the wider academic issues of security policy meant nothing to them, and that they had better get with it quickly. I think they are trying to, but it is not easy to change the system.[57]

In the past, the influence of service leaders has reinforced a general reluctance to face up to high level policy questions. During the postwar period, when Britain's political and economic position has been seriously undermined, the mindset of the services—committed like the armed services elsewhere to maintaining the importance of their particular defence role—has fused with unrealistic norms within the Foreign Office and the MoD about the kind of role which Britain ought to play in the world. None was anxious to face the question of whether the United Kingdom, which in 1986 had a per capita gross national product equivalent to that of the German Democratic Republic, could continue to maintain a wider spectrum of defence commitments than any other NATO member except France and the United States: the British nuclear force, the defence of the United Kingdom itself, the air/land contribution to the European mainland, the maritime operations in the Eastern Atlantic and the Channel, and 'out of area' commitments, principally now the Falklands.

How far does the failure to confront these issues reflect weaknesses in underlying strategic thinking on the part of military leaders? Greenwood, the leading advocate of a new defence review, puts the blame on procedural rather than intellectual factors. He argues that the managerial procedures of the MoD, which rely on *ad hoc* adjustments, surreptitious massaging of plans and spreading the misery equally by 'salami slicing', give power to the accountants, so that 'important shifts in priorities can occur without political or strategic debate taking place'.[58] But Julian Lider's study of British military thought, published in 1985 by the Swedish Institute for International Affairs, diagnosed major intellectual problems. Britain had no concept of a long-range military policy; military leaders and civilian strategists had failed to provide 'a thorough and truly authoritative analysis of the military challenges to Britain on which the military doctrine could be based'.[59]

The problem may have arisen because British military thinking has followed NATO thinking, and strategy has been neglected because NATO already has one—flexible response—even if it may be not so much a strategy as a political formula for avoiding strategic disagreement within the Alliance. But British strategists have tended to opt out of considering what alternative NATO strategies there could be, in order not to rock the boat. As Sir Frank Cooper put it:

> I think we have nationally—and this very much includes the Chiefs of Staff as well as the Government—opted out from saying, what kind of strategy should Britain follow. And we have simply tended to say, that is what NATO is for and is doing. But we have not got a policy. It is very embarrassing when you say so.[60]

The Royal United Services Institute, the professional association of the armed forces, has, perhaps in response to such perceptions, set up five study programmes to look at such issues as European security and the Atlantic Alliance, technology and defence procurement, and space and international security. These study programmes, externally financed by charitable organisations, financial institutions and industry, bring together practitioners, military and non-military, to evaluate the advantages and disadvantages of a range of policy options and act as a catalyst for debate.

It remains to be seen whether these programmes, albeit designed to be highly policy relevant and hence tending to operate within the existing parameters, will impinge in any significant way on defence decision makers. Military leaders are said to have had little influence on overall strategy in the post-war period. Lider, for instance, concludes that, even in those cases where military officers have attempted to inject a note of caution into the main lines of defence policy making, their influence has been small:

> Military thinkers stressed the need for modern, versatile and strong conventional forces, in Europe and overseas. They warned against building up nuclear forces at the expense of conventional ones. Nevertheless, military policy makers continued to emphasise the former.[61]

At the heart of this limited influence is the widespread belief that military hierarchies, in Britain as elsewhere, are incapable of seeing beyond the vested interests of their particular service to national defence priorities. To quote from Cooper again:

> Where I worry and I think that really it goes back into the roots of the organisation of the armed services themselves, is that if they have a destroyer, they want another destroyer, a better destroyer, a more expensive destroyer.[62]

Attempts to weaken the hold of the military over operational requirements have underlain most of the Ministry of Defence reorganisations since the war.[63] These reorganisations represent a sustained process of centralisation, to strengthen 'the increasingly dominant voice of the central defence organisation at all levels of defence planning from the Defence Minister downwards'.[64] Centralisation has involved, not only departmental reorganisation, but stricter control of public military comment on defence policy. Restricting what military leaders may say in public had the overt aim of preventing them from openly lobbying for sectional service interests. Even more convenient, though, is that such control prevents their publicly criticising government policy, for whatever reason.

A crucial episode in this respect was a lecture given at the Royal United Services Institute on 4 November 1959 by Lieutenant General Sir John Cowley, Master-General of the Ordnance. Cowley condemned the then Conservative Government's policy of heavy reliance on nuclear weapons to deter war, initiated in the defence white paper of 1957, and attacked what he considered to be the concomitant neglect of conventional forces. Cowley's public criticism of stated government policy brought a swift reaction. The Government announced in the House of Commons seven days later that in future all speeches by senior officers would have to be cleared by the Ministry of Defence. The Labour opposition supported the Government, thus ensuring that there would be no open and independent role for the services in any public debate on defence policy, whichever party was in power.[65]

Although this policy was slightly modified—military personnel, like civil servants, have for some 20 years been allowed to speak without attribution under 'Chatham House' rules at conferences and seminars run by certain universities and institutions—their freedom of speech is still strictly curtailed. The culture that has developed since the 1960s inhibits lively debate amongst the military in public. They are not supposed to address subjects that are potentially politically controversial, especially if they challenge the established defence parameters themselves rather than criticise within those parameters. In short, they must in no way embarrass the government. On the other hand, since defence has become a controversial party question, governments have increasingly made public use of military views when it serves not only governmental but possibly also party political interests.[66] Similarly, they have increasingly been tempted to edge over the fine line which separates the use of civil servants for government and for party ends. Michael Heseltine's creation of Defence Secretariat 19 (an anti-Campaign for Nuclear Disarmament propaganda unit staffed by civil servants) shortly after his arrival at the Ministry of Defence in January 1983, and its dissolution shortly after the general election of the following June[67] is the best known example.

The system, in which the services cannot publicly lobby for their own interests, has costs as well as possible benefits. Once they retire, it becomes evident that senior officers disagree about fundamental issues of defence policy.[68] But because active servicemen can only obliquely drop hints, those who disagree with government policy, for whatever reason, find it difficult to signal controversial issues to the public when policy is at a formative stage, and thus help informed public opinion to influence the outcome. For this reason among others the British defence debate is more inhibited than the American, operating as the latter does within a different constitution, deliberately constructed with checks and balances as a reaction against the British form of centralised decision-making. There Congress protects members of the services and of the bureaucracy who criticise government policy, but it has also been known, on occasion, to go on witch-hunts against officials or military leaders whose views it dislikes.

The muted military contribution to public defence debate was apparent in the case of Trident. Although some military leaders were uneasy about the choice made, the MoD was able to sustain a unanimous public front. A few retired officers—notably Lord Carver[69]—spoke out against the decision, but serving officers did not publicly express any doubts they might have had. The fact that officers are not supposed publicly to discuss defence policies still being formulated, to address subjects likely to be controversial between the political parties, nor to embarrass the Government, meant that opponents of the Trident option within the services could not lend their weight to a public debate *before* Government policy was announced, as would have happened in the United States, where 'Congress and the public at large can be kept informed of controversial issues when policy is at a formative stage and can thus exert an influence on the outcome'.[70] Even after the event, dissent was muted. The unexpectedly early retirement of Sir James Eberle, Commander-in-Chief, Naval Home Command, in 1983—a move which may not have been entirely unconnected with his lack of enthusiasm for Trident—was in characteristic contrast with the customary obstreperousness of French military leaders.[71]

The restrictions on public military comment have not, of course, prevented individual service lobbying, especially of Members of Parliament, and the system by which available money is split three ways, according to 'Buggins' turn', rather than according to agreed national defence priorities, remains sturdy in spite of repeated defence reorganisations designed to weaken it. Service leaders continue to subscribe, at least tacitly, to the principle of fair shares in the allocation of budgets. Their influence still combines with that of other interested parties—industrial groups such as the Society for British Aerospace Companies, trade unions with defence interests and Members of Parliament with commercial consultancies, defence industries or defence installations in their constituencies—to keep up overall levels of defence spending. This means that there has been unremitting pressure on ministers not to face up to budgetary problems—'never ever to take a decision because if no decisions are taken nobody is offended'.[72] Such combined pressure is still rarely resisted. The determined programme of naval cuts proposed in 1981–82 by Nott—a minister in the happy position of having no long term political ambitions and willing to return to merchant banking—was subsequently reversed.

The tacit alliance of military, political and industrial interests in Britain rests only, however, on resistance to overall defence cuts, not on agreement about which particular weapons should be procured. There is no military–industrial complex in the American sense: the military would almost always prefer to buy major items of equipment from the United States. In the past, given the commitment of all governments to maintain a national industrial defence base, they have not easily got their way, except in the case of strategic nuclear missile systems. In the case of Nimrod, for instance, the politico-industrial complex successfully put pressure on the Government to decide initially in favour of the British product, against the strongly held preferences of the Royal Air Force. Things may move more the services' way, in this respect, however, if there should in future be irresistible budgetary pressure to buy more from abroad. (See the preceding chapter.)

British Defence Decision Making

It is therefore possible to conclude that change is occurring only slowly in the services' contribution to defence decision making. The influence of service chiefs—less pervasive than that of senior civil servants—still consists mainly in the pressure successfully exerted over the overall size and inter-service division of the defence budget. It is only after retirement that a creative or unorthodox contribution to strategic thinking may be given full rein. By then, of course, those senior officers who contribute so trenchantly to the public debate are no longer part of the defence decision making system, but are part, along with pressure groups, of its environment. There, the processes of change are rather different.

The pressure groups

Parliament

Parliament, defined here as the first among pressure groups, has, like the decision making system itself, been slow to reflect the increasing prominence of defence as a public issue. Parliament has still only limited expertise in defence. Only a handful of Members of Parliament—perhaps two or three dozen in all—specialise in defence questions. That is probably a hangover from the days before 1983 when defence was not perceived as a vote-winner in Britain nor as a field in which Members could hope to make their reputations.

The defence specialists include members of the Defence Committee, described by one of them as 'the finest university in the world on defence matters', and the British representatives on the North Atlantic Assembly, 'a superb vantage point for knowledge of foreign defence ministries, for travelling and the acquisition of knowledge'.[73] A handful of these MPs operate virtually as defence professionals, belong to the main defence and foreign policy institutes in Europe and North America, participate in international conferences and seminars and have a string of publications to their name.[74] But although they may rub shoulders with the decision makers, it is doubtful whether their thinking makes any real input to the decision making process unless, as rarely happens, they are appointed to a defence portfolio.

A few Members of Parliament, such as Robin Cook, Kevin McNamara, Martin O'Neill (Labour) and Dafydd Elis Thomas (Plaid Cymru)[75] have developed an alternative defence expertise, critical of the major role of nuclear weapons and of military bloc politics, deriving from roots in the Campaign for Nuclear Disarmament, the schools of peace studies[76] and unofficial groups like the Alternative Defence Commission. They have been reinforced by some new MPs of the 1987 intake, such as Joan Ruddock, who rose to prominence through the anti-nuclear movement. But they have so far been excluded from any formal connection and have little informal contact with policy and decision making. They have yet to have experience of office or the management of defence.

The continuing weakness of House of Commons thinking on defence questions, compared with its handling of other policy areas or even, some would say, with the House of Lords' handling of defence issues, is manifest in the defence debates. There are several factors at work here. Internal discussion of defence questions is not popular within the major parties, except the Social Democratic Party, which was partly formed for defence policy reasons. Defence is a divisive issue for both the Labour Party and the Social and Liberal Democrats. Most of the small number of Labour Members of Parliament who have long specialised in defence questions are at odds with official party policy, and there is said to be little dialogue on defence between the two wings of the party.

The Conservative party, traditionally the party of 'strong defence', has other difficulties. A number of Conservative Members have a service background, but it is a myth that this, of itself, fits them for defence policy making. They may have little expertise outside a narrow area and may display a partisan loyalty to one particular armed service. As Nott sees it:

> There are a number of Conservative backbenchers who are knowledgeable about defence in one sense but in the sense of understanding the complexity of the issues and how they all come together, there are none outside the system [of ministers] because the 'defence buffs' as I used to call them, those who believe in defence, they are people who believe in spending more money, they want more tanks, more aircraft, more ships, more soldiers, sailors and airmen, they want a strong Britain. They are mainly a pretty gung-ho lot and they are subjected to the lobbying of the services who want more of everything. But the problems of defence in times of peace are virtually the problems of resource allocation.[77]

The traditional Parliamentary watchdog over government expenditure, the Public Accounts Committee, regularly scrutinises major items of defence procurement, often critically. The Public Accounts Committee remains more powerful than the newer House of Commons Defence Committee, but the amount of time which it can devote specifically to defence is limited, and it suffers, like other areas of parliamentary activity, from low levels of support staff. However, the Defence Committee, for its part, has steadily become more professional. When it was first formed, the Committee was influenced by what one member calls 'the strong anti-intellectual tradition of Parliament' but since then its overworked staff has been reinforced by outside academic or quasi-academic advisers (Lawrence Freedman of King's College, Keith Hartley of York University and the late Jonathan Alford of the International Institute for Strategic Studies) as well as retired military men such as Rear Admiral Gueritz. It does draw in extra specialist help for particular issues. But, apart from the House of Commons library, it is not able regularly to draw on any expert organisation working for Parliament itself, such as the American Office of Technology Assessment, which Congressmen can call on to disentangle knotty technical issues. Although the French Government created in 1983 a Parliamentary Office for the Evaluation of Scientific and Technological Choices, explicitly based on the American model, the Thatcher Government has resisted calls to follow suit, presumably feeling that the Government has little to gain by paying money to give Parliament its own independent sources of knowledge and advice.

The Defence Committee has managed to extract more information than hitherto from the Government and has in this sense made it more accountable. However, its effectiveness is limited by the long-standing practice of ministers and civil servants of giving as little information as they can get away with. The Committee expressed dissatisfaction with such attitudes during its 1984–85 hearings on defence equipment:

> We wanted to examine in-service dates, operational life, scale and phasing of expenditure. Our purpose was frustrated by vague and evasive answers and elegant but unhelpful hypothesis. Our experience in previous enquiries has not led us to expect the Ministry to volunteer information on matters which may be politically charged or potentially embarrassing; but we do expect proper answers to questions which are asked as part of our task of examining the 'administration, policy and expenditure' of the Ministry of Defence.[78]

A dependence on ministerial cooperation also limits the usefulness of parliamentary questions as a means of checking on the performance of the Ministry of Defence. Not only is the general run of Members lacking in ministerial experience which would help them to know what questions to ask, but the same is often true even of members of the Defence Committee.

For all this, the Committee has succeeded in bringing more retrospective information about defence than ever before into the public domain, and has secured information more readily from George Younger than from his predecessor as Secretary of State for Defence, Michael Heseltine. How far the Committee itself can interpret and analyse this information depends partly on how much time its part-time advisers are able to give, and partly on whether moves to generate funds for a permanent research base and to create a model of defence procurement, analogous to the Treasury model of the economy, are successful. At the moment, outside analysts can do more with the information which the Committee obtains than it can itself.

The Defence Committee is monitoring one or two major procurement programmes, including Trident, and asks about all the programmes at the time of the publication of the annual defence statement. Like the

Public Accounts Committee, it can act only retrospectively. And like Parliament as a whole, it tends to focus on time-scales within the five-year maximum life of a parliament, whereas major procurement programmes, as we have seen, may take double that time. Its monitoring of defence procurement is patchy; it does not have the resources to pay attention to everything. It has, in particular, still not been able to set up the much needed early warning system to detect signs of things going wrong in procurement programmes. It would, in any case, be difficult for a group of MPs to get a defence contract cancelled. But the Committee has helped to uncover the way procurement decisions are made, especially in the Trident case, and has scrutinised ministerial and civil service behaviour to an unprecedented degree, particularly in the Westland affair.

The Committee's effectiveness depends upon its being able to maintain the remarkable degree of inter-party co-operation, sustained during the period of the first and the second Thatcher Governments by its unlikely combination of left-wing Conservative and right-wing Labour members. By dint of this, it has managed to influence second-order decision making—that is, how programmes are managed rather than what programmes should be adopted—and other areas where there is no division along party lines. The Committee has been influential in encouraging reserve forces, in drawing attention to the implications of the decline of the merchant marine, and in these and other areas has forced issues to a head within the Government and compelled it to articulate its policies more carefully. It remains to be seen how the recent (1987) appointment to the Select Committee of two members of the Campaign for Nuclear Disarmament, John Evans and John McWilliam, will affect its work. On the one hand, these appointments will make the Committee more representative of the range of opinion on nuclear issues. On the other hand, they could reduce the Committee's effectiveness. The MoD could use the appointments as an excuse to deny the Committee access to secret documents, and defence ministries of other NATO governments could regard it with greater suspicion. Moreover, these new members would make it harder for the Committee to establish a united position, even in areas where there may prove to be no substantial policy differences between the parties, such as on conventional forces in Europe.

However, the Committee has always faced formidable problems in attempting to influence higher order or controversial defence decision making, where opinion is divided on party lines. Its role, and that of Parliament as a whole, in the Trident decision-making process, was characteristically negligible. Both the Labour and the Conservative Government tried to exclude Parliament from any involvement in the

Polaris-replacement decision, and frowned on Parliamentary discussion of the question before the Government itself announced its decision. Fred Mulley, the then Labour Secretary of State for Defence, tried unsuccessfully in the 1978–79 session to prevent the House of Commons Expenditure Committee from investigating the question, by means of an inquiry involving submissions from individuals and organisations with military or strategic expertise. In April 1980 an Early Day Motion was tabled in Parliament calling on the then Conservative Government to publish a green paper setting out the choices available to Great Britain, were the Polaris force to be replaced. But the Secretary of State said in reply that he did not accept that the idea of a green paper was either 'sensible or appropriate' and that the correct constitutional procedure was for the Government to take their decision and then explain and defend it before Parliament.[79] The newly formed House of Commons Defence Committee did indeed start an enquiry in June 1980 and planned to take evidence from the Ministry of Defence and from outside policy institutes and individuals. But just one month later, on 15 July 1980, the Government announced in the House of Commons its decision to purchase the Trident C4 system and invited the House to endorse this decision, before the Committee's enquiry was anything like complete.

The British Government was treading a path well-trodden elsewhere in Western Europe, where parliaments frequently receive information too late to exert influence on defence decision making. As the report to the Western European Union observed in 1979, 'Besides the quantity and the quality of the information necessary for parliamentary control, the time factor is just as important: that is the moment during the defence equipment procurement in which information is made available. Many defence committees of European parliaments seem to deal with these questions only *post facto*, even though it is very important to receive information in the preliminary stages'.[80]

A minority report of the Defence Committee noted that it 'saw no reason for action by the House before the Committee reported and consider the Government's actions in this respect to be less than courteous to both the House and its Committee'.[81] On 3 March 1981 the Secretary of State for Defence's motion to endorse the choice of Trident was carried by 316 votes to 248. But in any event, the Committee had decided at an early stage, voting along party lines, that the issue of principle about whether there should be a replacement for Polaris was essentially a political one, best left to the House of Commons as a whole and confined itself to looking into alternative forms of replacement and their opportunity costs. The minority report did, however, make a general comment on the way the decision had been arrived at:

> ... it will clearly be seen that Parliament's role in the decision to procure a successor system to Polaris has been limited to endorsing a decision already taken. Decisions on defence, and on Britain's strategic nuclear deterrent, have historically been taken by a small elite of very senior Cabinet Ministers, Civil Servants and Service Chiefs, and this present decision was certainly no exception.[82]

The minority report urged the then Government, and any future Government, 'to take Parliament, the public, industry, and the defence policy institutions more into its confidence in future'.[83]

Parliament was, as one would expect from existing constitutional traditions, largely restricted to the role of *post hoc* investigator, rather than participant, in the Trident decision. Since that time, the changes which have taken place have served to increase the accountability of government, but not to dent its monopoly of decision making. The achievements of the House of Commons Defence Committee notwithstanding, change in Britain has been belated compared with the United States, where as long ago as 1980 Judith Long and Franklin Reppy concluded that 'Congress has made in the past two decades considerable progress in transforming itself into a more effective partner in setting defense policy'.[84]

Defence academics and policy analysts

The area of defence and security studies in Britain has seen nothing like the phenomenal expansion which has occurred in the United States since the Second World War. However, the comparatively tiny field in Britain has been expanding in scope and prominence, particularly over the past 10 years, as defence has become a major public issue. The distance which Oxford and Cambridge have traditionally maintained from policy-oriented studies as the province of a few individuals with international reputations, such as Michael Howard and the late Hedley Bull, has changed somewhat, with the advent of the Cambridge Disarmament Seminar and the Oxford University Strategic Studies Group. Some 10 or so other universities and a small number of polytechnics have initiated policy-oriented work. The older disciplines of strategic studies, war studies and military history have been supplemented by the newer ones of peace studies and conflict resolution, specifically by the Bradford School of Peace Studies, and the Richardson Institute for Peace and Conflict Research at Lancaster University. Since the early 1970s, when Professor Michael Howard's courses on military history and strategy at Oxford were largely peopled by Americans, the number of British students of defence issues has slowly but steadily grown. The Open University's innovative course, *Nuclear Weapons: Inquiry, Analysis and Debate*, has not only attracted hundreds of undergraduates but includes radio and television programmes transmitted by the BBC to the public at large.

Simultaneously, a number of special interest groups have been founded within the universities to conduct research and/or disseminate information to the public. These include the Centre for the Study of Arms Control and International Security at Lancaster, founded in 1979; the Centre for Defence Studies at Aberdeen, founded in 1976; the Armament and Disarmament Information Unit within the Science Policy Research Unit at the University of Sussex, founded in 1978; and the Centre for International Policy Studies at the University of Southampton, founded in 1984.

These developments have undoubtedly enriched the quality of the public debate on defence and security issues, an area which some academics see as their primary role. The British public is probably better informed on defence issues now than at any previous time, but whether this will have any appreciable effect on official policy making is still an open question. Has this vigorous development succeeded in penetrating the traditionally closed system of defence decision making? How is the policy output of these institutions regarded?

A study of the European security and defence research sector, published in Belgium by Robert Rudney and Luc Reychler in 1986, concluded that there were three continuing problems inhibiting security and defence research in Britain: the highly centralised system of policy making; the Official Secrets Act; and problems of securing financial support.[85]

Cooper, typically of many civil servants past and present, tends to be dismissive of the policy contribution of defence academics, more mindful perhaps of their weakness relative to the American defence community than of their strength relative to that of most other European countries. He agrees that there are budgetary problems, but sees also serious methodological ones:

> In this country there is not a strong tradition particularly of numerical analysis, and I think a lot of defence analysts are historians *manqué* rather than what I would regard as a defence analyst. This has had the unfortunate effect that whereas there are quite a number of people in a variety of universities who will discuss concepts, when it comes down to what I call the nitty gritty, there has not been a great strength.[86]

What Cooper complains of is the way British defence academics, whose subject rides most often on the back of international relations, continue to concentrate on conceptual thinking and higher level policy questions. He would like to see them pay more attention to the technical and quantitative aspects of defence procurement, which form such a major item of public expenditure, pose intractable problems and urgently require 'an intellectual background of explanation'. What continues to prevent the kind of operational analysis which Cooper would like to see, and which is a strong feature of American defence analysis, is the lack of detailed information in Britain, the workings of the Official Secrets Act and the demands of commercial confidentiality. As things stand, academics cannot seriously contribute to operational analysis unless the MoD were to farm out such work to academic think-tanks, with their own secure areas and security cleared personnel, as the United States does with organisations such as RAND. Such a change of policy would involve long-term investment. The scope for any kind of quantitative defence analysis outside the defence establishment in Britain is strictly limited: with distinguished exceptions (such as Greenwood, who has for many years presented a critical challenge to Ministry costings and projections for the defence budget) defence academics have had necessarily to concentrate on higher order policy questions. Although, as we have seen, increasing amounts of information have become available retrospectively in the 1980s—there are, for instance, a number of detailed academic studies of Trident—the kind of information needed to support operational analysis is not usually available to academics. Even for broader policy studies, the information needed is not necessarily available in time—as far as its contributing to discussion before decisions are made is concerned—or easily or cheaply. The use of the Official Secrets Act as a political, rather than as a security instrument, means that the information needed for policy studies is often incomplete. This deficiency can only partially be rectified by the common academic practice of scanning the American archives and obtaining data under the American Freedom of Information Act on matters where the information is classified at home.

The persistent emphasis on policy rather than operational questions and the predominance of historians in the field are not altogether disadvantages. Certainly, British strengths in this field have always been history and politics; there have always been fewer lawyers in the field here than in the United States or continental Europe. Whereas the distinctive contribution of lawyers is their facility at asking good questions, what historians bring to the debate is scholarship, and an emphasis on international comparison. The value of this kind of approach is becoming more appreciated in the United States as exemplified by the journal *International Security*.

Rather than higher level conceptual questions being overdone by the British defence community, it might rather be said that they have still not been done enough. Defence policy is intimately linked not only to foreign policy and the structure of alliances but to national self image; not only to domestic industrial, employment and technology policy questions but also to issues of internal policing and of class and a whole range of psychological factors. The ideological implications of defence policy, the way different kinds of defence policy reinforce particular kinds of social, economic and political structures and underpin particular kinds of values, is only now beginning to be explored. Rather than concentrating too much on conceptual and value questions of this kind, mainstream defence academics have, on the contrary, continued to shy away from explicit discussion of the ideological dimensions of policy and the interconnections between defence and other major aspects of national policy.

At the same time, the number of technical or numerical analyses could also be increased. Where they have been carried out, often by people who became involved through the peace movement, they have had substantial impact on the public debate.

Apart from the persistent problems of secrecy and methodology, the relationship between defence academics and policy makers continues to be more problematic in Britain than in the United States. The Ministry of Defence, for its part, has traditionally distinguished between the scientific work which it sponsors in universities, and any involvement in policy questions. As Hussein sees it:

> Those scientists who are recipients of Ministry of Defence money for scientific research do not have a great deal of awareness of defence policy or why they are in fact undertaking the research other than the scientific reason, and if they did have serious interests in defence policy they would be very unlikely to be recipients of the money.[87]

However, the Ministry of Defence has attempted for some time to commission outside policy work. Clive Ponting, who left the MoD in 1985, recalls such exercises as a cynical bit of game playing:

> Closed government devalues the contributions of 'outsiders' in the decision taking process. I was able to observe a clear case of this at work inside MoD. Considerable 'unofficial' help was given to a number of outside institutes that specialise in studying strategic and defence problems such as the International Institute for Strategic Studies, Chatham House or the much less prestigious Royal United Services Institute. They produced a number of studies of various strategic problems but all based on publicly available information. The published studies were usually greeted with amused contempt within MoD. They were 'useful' if they took the existing departmental line but they were never taken into account during the decision taking process. Occasionally the idea would be floated that MoD should commission a study from one of the outside bodies. This was usually dismissed because they did not know enough to join in the 'advanced' level of debate inside the Ministry. Why did they not know enough? Because MoD would not release any additional information. Why not? Because it was classified. The only result of this policy was that debate about policy options was restricted to the vested interest groups inside MoD (the three services in particular) and no alternatives were put forward. The outside bodies become frustrated because they are ignored and ignored because they are frustrated.[88]

However, the MoD has made greater efforts to develop contacts with defence academics as defence issues have become more publicly contentious. As the 1986 *Statement on the Defence Estimates* put it:

> Informed public debate on questions of defence policy is of great importance in a democracy and the academic community has a major contribution to make. We are therefore keen to encourage closer links between the Ministry of Defence and the academic community in the field of defence studies. During the past year we have made a concerted effort to extend and improve these links and we are encouraging the universities and relevant institutes to play a more active role in the development of defence policy.[88]

The Secretariat of Policy Studies was set up under the 1984 reorganisation with a specific remit to improve and extend the MoD's links with the academic community. Chairmanship of the old Academic Studies Steering Group (now renamed the Defence Studies Steering Group) was moved from the Personnel and Logistics to the Policy area, precisely in order to shift the emphasis from training matters to the policy area. During the two and a half years of its existence, this Secretariat has considerably strengthened the MoD's relations with the academic world, both through personal contact and by providing an easy channel through which academics can reach the rest of the department. The long-established Defence Lectureship Scheme, in which the MoD funds defence lecturer posts in a number of universities, on the understanding that the costs of such posts will eventually be taken over by the university in question, underwent a significant revival, with a return to six lectureships as in the early days of the scheme. In 1986 the Secretariat instituted an annual series of contacts with the academic world, to which a range of people, including journalists, were invited, in the hope of getting some new ideas or at the least sharpening up old ones.

In 1983 the MoD began issuing research contracts to academics for policy work; and by 1985–86 four such contracts were in operation. Not unexpectedly, given the long-standing prejudice within the MoD against outside work, not all of the early resulting studies were thought to be valuable. From the civil service point of view, academics, lacking much experience of being on the 'inside', do not necessarily write in a form succinct enough for busy bureaucrats and in a manner which is 'policy relevant'. Most academics appear unfamiliar with the kind of problem-solving analysis which interests civil servants, and much of their expertise is too generalist for the kind of work which the MoD would like to commission. From an academic point of view, civil servants often seem to find it difficult to cope with contrary or divergent thinking, which challenges their assumptions or world view. For civil servants to rate highly the work of outside analysts, written without the benefit of classified information, is implicitly to devalue their own sources of power and authority.

Despite these difficulties, the MoD has not given up on its initiative. It is currently planning a trial scheme which will involve the creation of a mixed part-time research group of academics and MoD representatives, which it is hoped will be a fruitful way of combining the academics' expertise and broader perspectives with the civil servants' practical experience and knowledge of the system.

However, in any case, academics and policy makers in Britain do tend to have different needs and different priorities. The increasing demands of some defence academics' own universities may preclude them from doing work in the policy field. Some academics, for their part, remain wary of subjecting themselves to the increased ideological pressures which they see as inseparable from taking a government job in the highly centralised British defence establishment. The more pluralistic American system, in which outsiders are recruited in large numbers into government posts as administrations change, may be envied but it is not universally admired: the lure of a government post may, as Freedman observes, colour the whole orientation of an aspiring academic's output.[90]

The number of British defence academics and analysts with experience of the 'inside' remains very limited, and their contribution correspondingly distinctive. British defence academics—those in strategic studies as well as those in peace studies and conflict resolution—are not only separated from defence policy makers: their links are not as close as they might be with the military who figure so largely in their subject of study. Hussein sees this as a continuing weakness:

> The academic community does not have a very good feel for either policy formulation or for what the military services are about. I am not saying that anyone who does security studies has got to go on exercises and be a member of the Territorial Army but it does help if you actually understand what the services are at a practical level. That is, if you deal with people who are making a career then the closer you are, the more you understand what their thinking is about, the more likely your work is going to be relevant and the more likely you are to be able to pinpoint the problems specifically. So the community lives in rather an isolated fashion, both from the services' view and from [that of] the policy making organisation. At one end of the spectrum you have people who teach at Sandhurst and are members of the TA. At the other, you have people talking about nuclear policy and accidental nuclear war and they have not a clue about how anything works at all. I mean they have never seen it. Worse than that, they have no inclination to have a look. And the Ministry of Defence has very little inclination to show them.[91]

However, both the Navy and the Army, for their part, now claim to be giving higher priority to their relations with the academic world. They have, for instance, followed the example of the RAF and appointed a Head of Defence Studies specifically to liaise with the universities and specialist institutes.

The relationship between the London-based defence policy institutes and the policy makers remains less close than that between their counterparts in Washington, and this is very much the choice of both parties. The British-based policy institutes are anxious not to jeopardise their independence by becoming too closely tied to government. Although the International Institute for Strategic Studies, based in London but devoting most of its energies to international rather than British issues, accepted a small donation from the Ministry of Defence for its Foundation Appeal, it is keen to maintain its independence of the British Government, and never to be a formal government adviser. For this reason it has deliberately chosen not to seek government contracts for classified work, in contrast to, say, the RAND Corporation and others in the United States. The Royal Institute of International Affairs, concentrating on the political aspects of British security affairs rather than the military ones which are the province of the Royal United Services Institute, has closer links with government. There is a modified kind of 'revolving door', since two Foreign Office Fellows undertake research there each year, and the Institute sponsors an annual research seminar jointly with the Ministry of Defence. As we have seen, it was a seminar by Ian Smart, the Institute's Deputy Director and former Foreign Office civil servant, which began the academic public debate about Polaris in 1978, to the considerable inconvenience of the then Government. Although in that case Smart's conclusions were the same as those eventually reached by the official study groups, in general the Institute sees itself as in the business of examining the policy options which the Government prefers to reject. It has what its director calls a relationship of 'constructive tension' with the Government. It is part of its self-imposed task to ask the questions which the Government would prefer not to have asked and to give the answers which it would prefer not to hear.[92] Critics, mindful that ideological pressures are subtle and pervasive in this field, suggest that uncomfortable answers are rarely given, and then only in a whisper.

Typically, as we have seen, neither the policy institutes nor any of the university research centres were involved in the studies for the replacement of Polaris. Greenwood, criticising in his evidence to the Defence Committee what he believed to be the Ministry's neglect of the employment and other industrial implications of the decision, made the point that:

> by this juncture the American Department of Defense would almost certainly have committed several thousand dollars to relevant research by either a specialist contractor or one of the non-profit 'think-tanks' in the United States. In principle at least defence decision-making in the United Kingdom would be better informed if the Ministry of Defence, some economic department or even the Committee were to make similar use of the analytical resources available in this country.[93]

There are, one must conclude, some signs that the small élite circle of British policy makers has become more accessible to the academic defence community in the 1980s. Of course, many defence academics do not see it as their most useful or legitimate function to attempt to influence British defence policy. They do not necessarily concentrate on British defence issues at all, particularly in fields such as accidental nuclear war, where the most appropriate focus is the superpowers or NATO and the Warsaw Pact. Some prefer to direct their research and writings towards more powerful policy makers across the Atlantic. There remains, indeed, a continuing emphasis in British defence studies—as in British defence policy itself—on the Atlantic connection, rather than on the European dimension of security and defence issues.[94] British defence academics, like British civil servants, typically have a much closer professional dialogue with their English-speaking peers abroad, than with, for instance, the French or the Germans.

A number of academics are increasingly involved in the framing of party defence policies in Britain. All the major parties maintain defence working parties, usually supported by a small research staff, and the number and range of outside defence experts involved has grown as defence has moved closer to the centre of the political agenda. The Conservative Party is thought to be more sceptical of the value of academics—'intellectuals'—feeling more comfortable with the military and officials.[95] The Labour Party is less comfortable with military officers, and it is in any case difficult for serving officers, subject as they are to Chatham House rules, to have the kinds of contact with British opposition parties which some of the German military have with the German Social Democratic Party. But the Labour Party has a substantial contingent of alternative defence academics who provide informal briefings and who occasionally serve on its working parties. The SDP bristles with more mainstream strategists.

But academic/party links, like academic/government links, are still seen as problematic in Britain. On the one hand, it is only when academic ideas begin to gain a political constituency that they can start to have an effect. On the other hand, academics may feel that they risk compromising not only their academic credibility but their intellectual honesty—not to mention their sources of information—if they stake out too clearly a defined political position. As Lawrence Freedman sees it:

> It is one thing to have a broad political sympathy but you've got to be in a position to subvert your own movement by sticking with a bit of analysis that comes to a rather awkward conclusion, and not trying to put forward these package deals in which you prove that everything is for the best in the best of all possible worlds.[96]

Conclusion

Despite the changes which have taken place in Britain with the ending of the inter-party defence consensus and the rise of defence as an issue of political importance, only limited changes have taken place in the processes of defence decision making in Britain. In spite of the achievements of the House of Commons Defence Committee in bringing information into the public domain and making the Government more accountable, the centralised decision making enshrined in the British constitution ensures that policy making remains the responsibility of a small élite. Although the public defence debate is better informed and livelier than ever before, it is still only running parallel to the decision-making process and only occasionally exerting a slight impact on it. In spite of notable attempts by the MoD to increase its links with academic and other expertise outside, the best available talent is still not mobilised in Britain as it is in many other countries to help seek out answers to the many complex issues on which defence decisions need to be taken.

One of the undoubted benefits of the end of the defence consensus and the opening up of public controversy has been that more information than ever before in Britain is now in the public domain. It would be wrong to underestimate the amount of information which can now be obtained by tireless, time-consuming and possibly expensive investigation. However, that does not mean that there is no continuing problem with secrecy. The question of when information becomes available is crucial: all too often such information becomes available only in time to support an historical analysis, rather than to enable outsiders usefully to contribute to a debate informing decision making. The Official Secrets Act is too often used to serve party political rather than legitimate security interests. Key decisions such as that on Trident continue to be made by small groups of men operating in secret. Robert Oppenheimer's strictures on the way in which the H-bomb decision was arrived at in the United States in the early 1950s remain applicable to the Britain of the 1980s:

> There is grave danger for us in that these decisions have been taken on the basis of facts held secret. This is not because the men who must contribute to the decisions, or must make them, are lacking in wisdom; it is because wisdom itself cannot flourish, not even truth be determined, without the give and take of debate or criticism. The relevant facts could be of little help to an enemy; yet they are indispensable for an understanding of questions of policy.[97]

This pattern will not easily be broken: it has lasted so long because particular interests are served by its continuance. Knowledge is power and will not easily be surrendered. Some legitimate public interests are said to be protected by the present levels of secrecy. The business of defence decision making, with its imponderables and mismatches, is highly complex without the added complications of public political controversy and overt industrial lobbying. Public participation and the extensive use of academic consultancy in defence decision making may be both expensive and wasteful, as the United States bears witness. Structures set up to facilitate the political processes of participation may conflict with, and will certainly complicate, managerial processes. There is always the risk that the main beneficiaries of greater openness will be industrial lobbies or other vested interests.

However, in the long run the costs of exclusivity exceed the costs of participation. A fundamental rethinking of the basis on which information is classified, for instance, is costly only in terms of political expediency. As a guiding principle, information relevant to policy questions should be in the public domain, whereas information relevant to operational questions need not necessarily be.[98] The convenient practice by which politicians of all parties raise the cry of security for political ends has to be resisted.

The exclusivity of defence decision making in the past, and the compartmental structures within which it has taken place, have had associated costs for the kind of thinking about defence which has taken place. So powerful is the decision makers' shared interpretation of experience and tacitly agreed strategies, norms and assumptions, that fundamental issues of defence policy remain not only undiscussed but undiscussable. While the French and the Germans, for instance, have been laying the foundations for a radical rethinking of their defence co-operation in Europe, British policy makers remain locked into the English-speaking dialogue which underpins the special relationship with the United States, despite much talk of European defence co-operation at the margins.

Given the scale and complexity of the defence dilemmas which lie ahead, the defence establishment might usefully see the livelier, better informed and more controversial public environment for defence decision making as a resource, rather than a threat. But there is little sign of that yet.

Notes

1. Quoted in Adam Roberts, *Nations in Arms*, Chatto & Windus, for the International Institute for Strategic Studies, 1976, p. 250.
2. Prominent among these were the Campaign for Nuclear Disarmament, which started regular lobbying and briefing about 1982 and providing regular public information the following year; Scientists Against Nuclear Weapons (1981) and its offshoot, the Verification Technology Information Centre (1986); Defence Information Groups (1983); the London Centre for International Peacebuilding (1983) and the associated Generals for Peace and Disarmament; and the Oxford Research Group (1982).

3. Such as the Institute for European Defence and Strategic Studies; the Coalition through Peace and Security; and the Institute for the Study of Conflict.
4. Lauren H. Holland and Robert A. Hoover, *The MX Decision*, Westview Press, 1985, p. 254.
5. Assembly of the Western European Union, *Parliaments and Defence Procurement*, Document 807, 22 May 1979, pp. 3, 18.
6. Ibid., p. 8.
7. Michael D. Hobkirk, *The Politics of Defence Budgeting: a Study of Organization and Resource Allocation in the United Kingdom and the United States*, National Defense University Press, 1983, p. 79.
8. Anthony Eden, *Full Circle*, Cassell, 1960, p. 367: 'The Prime Minister is ultimately responsible for all important decisions on defence. That is how it should be.'
9. Eden was Minister of Defence from May to December 1940, Attlee from July 1945 to December 1946, and Macmillan from October 1954 to April 1955.
10. Ian Beckett and John Gooch, *Politicians and Defence*, Manchester University Press, 1981, p. viii.
11. Max Hastings, *Sunday Times*, 19 January 1986.
12. An exception was Alun Gwynne Jones, defence correspondent of *The Times*, created Baron Chalfont in 1964 and appointed Minister of State at the Foreign and Commonwealth Office.
13. C. Hernu, *Soldat-Citoyen: Essai sur la Défense et la Sécurité de la France*, Flammarion, 1975; *Chroniques d'Attente*, Tema, 1977; *Nous... les Grands*, Boursier, 1980.
14. Beckett and Gooch, see note 10, p. 118.
15. Interview with Sir John Nott, 9 April 1986.
16. Edwin A. Deagle, 'Organization and Process in Military R and D', in Franklin Long and Judith Reppy, *The Genesis of New Weapons*, pp. 163–4.
17. Hobkirk, see note 7, p. 118.
18. Senior defence civil servants are said to have invented a comic game in which there are four teams—A: the Army, who want kit; B: companies, who make kit and want money; C: the taxpayer who provides the money (the losers); D: the Ministry who take kit from B and give it to A and take money from C and give it to B; *Daily Telegraph*, 28 January 1986.
19. Sir Clive Whitmore, 'Ministry of Defence reorganisation: the implementation of change', *Journal of the Royal United Service Institute*, March 1985.
   The MoD takes: 50% of the output of the British aerospace industry; 60% of the output of the British ordnance industry; 20% of the output of the British electronics industry; 40% of the output of the British shipbuilding and ship repairing industry. *Defense and Economy World Report*, no. 1005, 2 June 1986.
20. Interview with Sir Frank Cooper, 9 April 1986.
21. Ponting, Assistant Secretary at the Ministry of Defence, was charged (and subsequently acquitted) in 1985 under Section 2 of the Official Secrets Act with disclosing official information about the sinking of the *General Belgrano* during the Falklands War.
22. Interview with Clive Ponting, 23 April 1986.
23. In August 1986 the Comptroller and Auditor General criticised the MoD for inadequacies and inefficiencies in its procurement programme for major items of defence equipment: it had failed to overcome problems of escalating costs and delays in major contracts. *Ministry of Defence: Control and Management of the Development of Major Equipment*, National Audit Office, HC Paper 568, 1985–86, HMSO, August 1986.
24. Cooper interview, see note 20: 'The Ministry of Defence is a very large conglomerate. It has got an ethos of its own which is different from other government

departments; it has got three services inside it; it has not merely got the old fashioned Ministry civil servants, it has got very talented engineers and scientists and, uniquely, it spends its own money.'
25. Interview with Sir James Eberle, 9 February 1987.
26. Nott interview, see note 15.
27. G. M. Dillon, *Dependence and Deterrence*, Gower, 1983, p. 125.
28. Argyris and D. Schon, 'What is an organisation that it may learn?', in R. Paton, S. Brown, R. Spear, J. Chapman, M. Floyd and J. Hamwee (eds.), *Organizations: Cases, Issues, Concepts*, Harper & Row, 1984.
29. David S. Yost, 'Radical change in French defence policy?', *Survival*, January/February 1986, p. 61.
30. 'Changing the establishment', lecture delivered at the London School of Economics, 1986. Cf. Robert Rudney and Luc Reychler, *In Search of European Security: an Assessment of the European Security and Defence Research Sector*, Leuven University Press, 1986, p. 273: 'Several sources mentioned the interest and encouragement of Sir Frank Cooper, the recently retired Permanent Under Secretary of Defence ... in opening up lines of communication'.
31. Laurence W. Martin, 'The market for strategic ideas in Britain: the "Sandys era"', *American Political Science Review*, vol. 56, part 1, March 1962.
32. Interview, Laurence Martin, 25 June 1986.
33. Cooper interview, see note 20.
34. The Union of Small Companies claims that between 1979 and 1983 1,809 members of the armed services and civil servants—1,404 from the MoD—applied for

business appointments and only 15 were rejected, *Daily Telegraph*, 8 January 1986.
35. Interview with Gloria Franklin, 21 July 1986.
36. A post created in 1964. Holders to date: 1965–66 Sir Solly Zuckerman; 1967–68 vacant; 1969–70 Sir William Cook; 1971–77 Professor Hermann Bondi; 1978–83 Professor Ronald Mason; 1983– Professor Richard Norman. All except Sir William Cook were external appointments.
37. Seminar given by Sir Ronald Mason at the Open University, 1986.
38. Interview with Sir Ronald Mason, 18 April 1986.
39. Quoted in Harvey Brooks, 'The scientific adviser', in Robert Gilpin and Christopher Wright (eds.), *Scientists and National Policy Making*, Columbia University Press, 1964, p. 80.
40. *UK Military R & D*, Report of a Working Party, Council for Science and Society, Oxford University Press, 1986, p. 20.
41. Some 50 British scientists and engineers went to the United States to work on the Manhattan project, at a time when most of the British-born scientists with high reputations were already on war work. Margaret Gowing, *Britain and Atomic Energy, 1939–1945*, Macmillan, 1964, p. 268.
42. Interview with Lawrence Freedman, 1 July 1986.
43. Mason interview, see note 38.
44. Ian Smart, *The Future of the British Nuclear Deterrent: Technical, Economic and Strategic Issues*, Royal Institute of International Affairs, 1977.
45. C. E. Lindblom, 'The science of muddling through', in D. S. Pugh (ed.), *Organization Theory*, Penguin Books, 2nd edition, 1984, p. 244: 'Evaluation and empirical analysis are intertwined; that is, one chooses among values and among policies at one and the same time ... one simultaneously chooses a policy to attain certain objectives and chooses the objectives themselves.'
46. On 11 March 1986 the Secretary of State for Defence announced a revised cost of £9,869 million for the Trident programme.

47. The Thatcher Government has said that it will not load up the Trident with more than 8 warheads per missile out of a possible maximum of 14.
48. Interview with Sir James Eberle, 9 February 1987. Eberle considers that the Chiefs of Staff fell into a trap.
49. House of Commons Defence Committee, *Strategic Nuclear Weapons Policy*, 4th Report, HC Paper 36, 1980–81, Q284.
50. Lindblom, see note 45, p. 247: 'It is a matter of common observation that in western democracies public administrators and policy analysts in general do largely limit their analyses to incremental or marginal differences in policy that are chosen to differ only incrementally.'
51. Interview with the late Colonel Jonathan Alford, 18 July 1986.
52. Interview with Sir James Eberle, 21 July 1986.
53. Rudney and Reychler, see note 30, p. 258.
54. Hew Strachan, 'The British Army and the study of war—a personal view', *Army Quarterly and Defence Journal*, vol. 111, no. 2, April 1981, p. 137.
55. Rudney and Reychler, see note 30, p. 279.
56. Interview with Farooq Hussein, 27 May 1986.
57. Eberle interview, see note 52.
58. David Greenwood, cited in John Baylis, 'Greenwoodery and British defence policy', *International Affairs*, vol. 62, no. 3, Summer 1986, p. 455.
59. Julian Lider, *British Military Thought after World War II*, Gower, 1985, p. 564–5, 567. 'Military thought' here includes that of the mainstream defence community, not just that of the military.
60. Cooper interview, see note 20.
61. Lider, see note 59, p. 206.
62. Cooper interview, see note 20.
63. Those took place in: 1945; 1959; 1964; 1982–3; and 1984.
64. Hobkirk, see note 7, p. 17.
65. Ibid., p. 58.
66. Such as the Government produced video cassette 'Keeping the Peace', issued in 1987.
67. Clive Ponting reports that the Permanent Under Secretary was unhappy about the use of civil servants in this instance; interview, see note 22.
68. See, for instance, John Baylis (ed.), *Alternative Approaches to British Defence Policy*, Macmillan, 1985, and contributions by former Chiefs of the Defence Staff, Field Marshal Lord Carver, Admiral of the Fleet Lord Hill-Norton, and Marshal of the Royal Air Force Lord Cameron.
69. Lord Carver, 'Why Britain should reject Trident', *Sunday Times*, 21 February 1982.
70. Hobkirk, see note 7, p. 107.
71. In 1985 General Philippe Arnold, commander of the first Tank Division, publicly denounced the long delays in the deployment of the modern battle tank; that same year General Lacaze, Chief of the Joint General Staffs, openly warned that resource problems were likely to 'compromise the operational capacity of the armed forces'. Jolyon Howorth, 'Resources and strategic choices: French defence policy at the crossroads', *World Today*, May 1986, p. 77.
72. Cooper interview, see note 20.
73. Interview with Bruce George, MP, 10 June 1986.
74. For example, Julian Critchley, Conservative MP for Aldershot, Vice Chairman of the Conservative Defence Committee; former Chairman of the Western European Union Defence Committee; member of the International Institute for Strategic Studies; publications include: *Collective Security* (with O. Pick), Macmillan, 1975; *NATO and the Soviet Union in the 80s*, Macmillan, 1980; *Warning and Response*, Cooper,

1978; *Cruise, Pershing and SS 20* (with J. Cartwright), Brasseys, 1985; and *Heseltine*, Deutsch, 1987.
75. Robin Cook, MP for Livingston, opposition spokesman on European and Community Affairs, 1983–85; Kevin McNamara, MP for Hull North, opposition spokesman on defence, 1982–83; on defence and disarmament, 1983–85; deputy opposition spokesman on defence 1985–87; members of the Parliamentary Assembly, NATO; Martin O'Neill, MP for Clackmannan; Dafydd Elis Thomas, MP for Meirionnydd Nant Conwy.
76. Bradford University School of Peace Studies; the Richardson Institute for Conflict and Peace Research, Lancaster University.
77. Nott interview, see note 15.
78. Quoted in: Council for Science and Society, see note 40, p. 60.
79. House of Commons Defence Committee, see note 49, p. xxxviii.
80. Western European Union, see note 5, p. 18.
81. House of Commons Defence Committee, see note 49, p. xxxix.
82. Ibid., p. xxxviii.
83. Ibid., p. xxxix.
84. Long and Reppy, see note 16, p. 192.
85. Rudney and Reychler, see note 30, pp. 280–1.
86. Cooper interview, see note 20.
87. Hussein interview, see note 56.
88. Clive Ponting, *The Right to Know*, Sphere, London, 1985, p. 208.
89. *Statement on the Defence Estimates 1986*, Cmnd 9763-I, 1986, p. 49.
90. Lawrence Freedman, 'The development of the think tank', *Journal of the Royal United Services Institute*, vol. 127, no. 1, March 1982, p. 16.
91. Hussein interview, see note 56.
92. Eberle interview, see note 52.
93. House of Commons Defence Committee, see note 49, p. 50 note.
94. Rudney and Reychler, see note 30, p. 283.
95. Kenneth Hunt, 'The British defence debate—a case study', *Journal of the Royal United Services Institute*, vol. 127, no. 2, June 1982, p. 9.
96. Freedman interview, see note 42.
97. Quoted in: Herbert York, *The Advisers: Oppenheimer, Teller and the Superbomb*, Freeman, 1976, p. 72.
98. Interview with Lawrence Freedman, 10 February 1987.

About the Authors

The authors were all, at the time this book was conceived, members of the Technology or Science faculties of the Open University.

Chris Bissell is Lecturer in Electronics in the Open University and author of *Control Engineering*, Van Nostrand Reinhold, 1988.

Margaret Blunden was Lecturer in Systems in the Open University from 1978 to 1987 and is now Head of the School of Social and Policy Studies of the Polytechnic of Central London. She is co-editor of *The Vickers Papers*, Open Systems Group, Harper & Row, 1984.

David Carlton is Senior Lecturer in History at the Polytechnic of North London; he was seconded to the Open University from 1983 to 1986. He is the author of *Anthony Eden: A Biography*, Allen Lane 1981 and is co-editor (with Carlo Schaerf) of *The Dynamics of the Arms Race*, Croom Helm (1975), *International Terrorism and World Security*, Croom Helm (1975), and *The Arms Race in the 1980s*, Macmillan (1982).

Owen Greene was Research Fellow in the Science Faculty of the Open University from 1983 to 1986 and is now Lecturer at the Bradford School of Peace Studies. He is author or co-author of many articles and a number of books on defence issues, including *London After the Bomb* (1982), *Europe's Folly: Facts and Arguments about Cruise* (1983), *Doomsday Britain after Nuclear Attack* (1983), *Without the Bomb* (1983), *Nuclear Winter: the Evidence and the Risk* (1985), and *The Politics of Alternative Defence* (1987).

David Lowry is Research Fellow in the Energy and Environment Research Unit of the Open University, co-editor of *Issues in the Sizewell B Inquiry*, Centre for Energy Studies, London, 1983 and editor of *Reflections on Britain's nuclear history in conversation with Lord Hinton*, Energy Research Group, Milton Keynes, 1984. He is the director of the European Proliferation Information Centre in London.

John Monk is Professor of Electronics (Digital Systems) in the Open University Faculty of Technology and works with the Advanced Networked Systems Architecture Project, a multi-company collaboration.

John Naughton is Senior Lecturer in Systems in the Open University and is an external adviser to the Manpower Services Commission on Artificial Intelligence.

Joyce Tait is Senior Lecturer in Systems in the Open University, contributing editor of the journal *Environment* and member of the JUPITER consortium on technology management.

Index

Acheson, D.   95
Acheson–Lilienthal report   129
Ackerman, Thomas   19–23
Acland-Hood, Mary   81, 85
Advanced gas cooled reactors (AGR)   128
Advanced Research Projects Agency (ARPA)   83
Advanced Technology Bomber (Stealth)   207
Agricultural production   25
Airborne warning and control aircraft (AWACS)   176
Air-launched cruise missile (ALCM)   106, 107
Alfonsin, Raoul   49
Alford, Jonathan   220, 228
Allaun, Frank   155
Allen, Richard   66
Allison, Graham   97
Alvarez, Luis   37, 93, 94
Ambio   18, 19, 43
AMD26LS20   173
American Philosophical Society   77
Anglo-American atomic relations   132–44
Anglo-American special nuclear relationship   128
Anti-Ballistic Missile (ABM) Treaty   5, 66, 67, 69, 70, 110
Anti-ballistic missiles (ABMs)   85, 99–101
Anti-nuclear movement   15, 20, 43
Arms Beyond Doubt; the Tyranny of Weapons Development   80
Atlas/Titan I   99
Atomic bombs   86
Atomic Energy Act   132, 133, 139
Atomic Energy Commission (AEC)   93
Atomic fuel   133
Atomic tests   93
Atoms for Peace programme   129, 130, 139, 145, 151
Australia   50

B-1 bomber   107
B-29 bomber   87, 93
B-52 bomber   78, 107
Bacon, Commodore R.   35
Baker, John   146
Barnaby, Frank   18
Barnham, Keith   140
Beilenson, Laurence   64, 65
Belt-way bandits   28, 32
Bendetsen, Karl R.   67
Benn, Tony   128, 154–6
Birk, J. W.   21
Birks, John   18
Blackett, P. M. S.   216
Blunt, Sir Anthony   159
Boiteux, Marcel   154
Bradbury, Norris E.   90
Bradley, General   96
Brezhnev, L.   47
British Telecommunications   186
Brooks, Harvey   85, 87, 88
Brown, Harold   177
Bryuson, Admiral Sir Lindsay   195
Bundy, McGeorge   67
Burke, Admiral   133

Calder Hall   131, 133, 140, 143
Campaign for European Nuclear Disarmament   51
Campaign for Nuclear Disarmament   43, 44, 51, 144, 148, 155, 158, 227, 230
CANDU   138
Carey, C. J.   110
Carter, Jimmy   61, 145
Carver, Lord   225
Casey, William   66
Central Electricity Generating Board (CEGB)   140–60
Central Intelligence Agency   83
Centre for the Consequences of Nuclear War   27
Chapel Cross   143
Chazov, E. I.   20, 22

247

Chernobyl accident 156, 158
Chevaline programme 10, 78, 89, 108–12, 124, 149, 206
Chief Scientific Advisers 216
China 33, 50
Chowcat, John 194
Christmas Tree syndrome 175
Churchill, Sir Winston 208
Citrine, Lord 135
Civil–military interactions 121–5
Climatic catastrophe 15, 44
Climatic consequences of nuclear war 18–22, 25, 48
Clwyd, Mrs 165
Cohen, Senator 38
Cohen, William S. 36
Coleraine, Lord 150
Committee of the National Research Council 23
Committee on the Present Danger 66
Conant, James 94
Conference on the World after Nuclear War 20, 26
Congressional Joint Economic Committee 176
Cook, Robin 227
Cooper, Sir Frank 211, 214, 215, 223, 224, 233
CORAL 66 195
Council for Science and Society 189
Counter-radar-jamming techniques 171–2
Courier, The 46
Cowley, Lieutenant General Sir John 224
Crossman, Richard 143
Cruise missiles 10, 43, 44, 62, 89, 128
Crutzen, Paul 18, 21

Dalyell, Tam 159
Dean, Gordon 94
Defence budgets 168
Defence contracts 82–3
Defence decision making 10, 205–45
 civil servants' role 212–20
 defence academics' role 232–9
 military role 220–7
 ministers' role 208–12
 Parliament's role 227–32
 policy analysts' role 232–9
 pressure groups' role 227–32
Defence policy making 121–5, 127, 167–204

Defence Technology Enterprises (DTE) 194, 195
Defender destroyer 169
Defense Advanced Research Projects Agency (DARPA) 5, 83
de Gaulle, General Charles 152, 208
de la Madrid, Miguel 49
DeLauer, Richard 9, 34, 61, 71
Delhi Declaration 49
Denton, Jeremiah 36
Department of Defense 82, 84
Department of Energy 83
Department of Trade and Industry 125
Dillon, G. M. 213
Dixon, Alan J. 36
Does Technology Policy Matter 191
Dombey, Norman 148
Dreadnought 138
Duff, Sir Anthony 218
Dulles, John Foster 132
Dungeness B power station 131
Duquesne Light Company 131
Durie, Sheila 148
Dust clouds 18–19, 25

Eberle, Admiral Sir James 222, 226
Economics of Defence Spending 189
Eden, Anthony 208
Ehrlich, Ann 44
Ehrlich, Paul 21, 22, 44
Eisenhower, President 3, 78, 82, 93, 129, 130, 134, 137, 145, 151
Eklund, Sigvard 145
Electrical Power Engineers' Association 148
Electromagnetic pulse (EMP) 174, 175
Electronic counter-counter-measures 171, 172
Electronic counter-measures 171, 172
Electronics
 British policy implementation 192–9
 British policy research 185–92
 civil and military 167–204
 trends and policy dilemmas 168–76
Electronics Economic Development Council (EDC) 181, 186, 193
Electronics industry, international comparisons 176–85
Energy policy 127, 128, 149–51
Ergas, Henry 178, 191
Ericsson, L. M. 195
EURATOM 138, 139, 152, 153, 157
Europe 14, 20

European Coal and Steel Community 152
European Economic Community 138, 152, 153
Evans, John 230
Exon, J. James 36

F-4 Phantom 170
F-16 fighter aircraft 6
F-16C combat aircraft 173
F-18 programme 170
F-104G Starfighter 169
Fairchild 178
Fallows, J. 6
Fault tolerance 173
Fermi, Enrico 91
Fishlock, David 147–8
Fletcher, James C. 72
Follow on forces attack (FOFA) 170
Ford, Gerald 61
Foreign Affairs 22, 28
France 33, 50, 151, 152, 153, 179, 180
Freedman, Lawrence 73, 228, 239
French, Denzil 143
Fuchs, Klaus 96
Fuelling the Nuclear Arms Race: the Links between Nuclear Power and Nuclear Weapons 148
Fusion technology 91–3

Galosh ABM system 108
Gamow, George 94
Gandhi, Rajiv 49
GEC-Plessey 198
General Advisory Committee (GAC) 93–6
General Belgrano 159
Gillespie, C. M. 36, 37, 39
Gilpin, R. 179
Giraud, André 209
Glenn, John 36, 37
Gol'danskii, Academician 46
Goldwater, Barry 36, 37
Golitsyn, Georgiy 44
Goodlad, Alistair 157, 160
Gorbachev, M. 45, 47, 88
Goure, Leon 36
Graham, Daniel O. 67, 68, 69
Grapple-1 Test 136
Gray, Colin S. 66
Greater London Trade Union Resource Unit 189
Greenpeace 50

Greenwood, David 219, 223, 238
Greenwood, Ted 97, 98, 101, 102
Greve, Frank 71, 72
Gromyko, Foreign Minister 45
Groves, General 113
Gueritz, Rear Admiral 229

H-bomb 79, 84, 87, 89–97, 136, 240
Hagelstein, Peter 87
Haig, Alexander 65
Hamilton, Archie 165
Hannah, L. 135, 140
Hart, Gary 36, 38
Hartley, Keith 228
Harwell, Mark 36, 38, 42
Hastings, Max 209
Healey, Denis 43, 208, 214
Heath, Edward 109, 153
Heritage Foundation 67, 69
Hernu, Charles 209
Heseltine, Michael 193, 194, 195, 196, 208, 225, 229
Hesketh, Ross 136, 138, 147
High Frontier 67, 69
High technology investment programmes 151–4
Hinckley Point 144
Hinslow, Carl 94
Hinton, Lord 140, 144, 147, 149–51
Hiroshima 18–19, 86, 87, 91
Hiss, Alger 95
Hobkirk, Michael 207
Hoffman, Fred S. 72
Holland, L. H. 103, 105
Hollis, Sir Roger 159
Hoover, R. A. 103, 105
House of Commons Defence Committee 12, 205, 207, 219, 220, 227–9, 231, 232, 240
House of Commons Public Accounts Committee 206, 228–30
Howard, Michael 232
Howe, Sir Geoffrey 78
Huisken, R. 107
Humphrey, Gordon J. 36
Huntingdon, Samuel 210
Hussein, Farooq 222
Hydrogen bomb 10, 89

ICBMs (intercontinental ballistic missiles) 64, 100, 104
Ikle, Fred C. 66, 71
Industrial policy 127

Industrial production 25
Information technology 167, 176
Intercontinental ballistic missiles
    (ICBMs) 62, 66, 67, 69, 70, 88
International Council of Scientific
    Unions 23, 27
International Institute for Strategic
    Studies 73
International Physicians for the
    Prevention of Nuclear War
    (IPPNW) 19
Iranian affair 1986–87 72
Izvestia 46

Jackson, Robert 193
Japan 180–2
Joint Chiefs of Staff (JCS) 70–1, 93
Joint Committee on Atomic Energy
    (JCAE) 93–5, 133, 134
Jones, Peter 109
Jones, R. V. 216

Kahn, Herman 29
Kaldor, Mary 175, 192
Kapitsa, Professor 46
Kearny, Cresson 31, 32, 45
Kennan, George 67, 94, 96
Keyworth, George A. II 70, 71, 72
Kirkpatrick, Jeane 66
Kissinger, Henry 107
Konopinsky, Emil 91
Korean War 130
Kruschev 47

Labour Party 43, 50
Laird, Melvin 107
Lapp, Ralph 80
Laser
    guided-missiles 172
    range-finders 172
    target-designators 172
    warning-systems 172
Lawrence, E. O. 93, 94
Lawson, Nigel 157, 158
Layfield, Sir Frank 156
Leopard 2 tank 169
Levene, Peter 197, 215
Levin, Carl 36
Levitt, M. S. 189
Lider, Julian 224
Light water reactor (LWR) 129, 141, 152

Lilienthal 94, 95, 96
Literary Gazette 46
Long, Judith 232
Los Alamos Laboratory 91
Lucas, Lord 198

M-16 rifle 6
M47 tank 169
McFarlane, Robert 70, 71, 72
McMahon, Senator Brien 93, 94, 95, 96, 142
Macmillan, Harold 136, 137, 139
McNamara, Kevin 227
McNamara, Robert S. 67, 101
McWilliam, John 230
Maddock, Sir Ieuan 186–9, 193, 195, 196
Magnox reactors 133–4, 136, 140, 142, 144, 148, 150, 151, 153, 154, 157, 160
Making of MIRV: A Study in Defense Decision Making 98
Malenkov 47
Malone, T. 34
Manhattan project 79, 113
Mark 12 system 99, 100
Marshall, Lord 138, 148
Martin, Laurence 214–15
Mason, Sir Ronald 214, 216
May, M. 34
Menaul, Air Vice-Marshal Stuart 111
Merchant shipping 127
Microelectronics 170, 182
Ministry of Defence (MoD) 12, 82, 186, 206, 207, 210–13, 215, 216–25, 230, 234–8, 240
Ministry of International Trade and Industry (MITI) 181
Minuteman 99, 102
Minuteman II 99, 100
Minuteman III 98, 100
MIRV (multiple independently targeted re-entry vehicles) 10, 70, 78, 79, 85, 87–102, 108–10, 217
Mondale, Walter 73
Moore, Captain J. E. 175
Moore, John 158
Moscow Peace Forum 48
Mountbatten, Admiral Lord 78, 133
Muggeridge, Malcolm 205
Mulley, Fred (*later* Lord Mulley) 208, 231

Multiple Launch Rocket System 171
Multiple re-entry vehicle (MRV) technology 98
Mutual assured destruction (MAD) 63, 68, 101, 102
Mutual Defense Agreement 124, 139–43, 145, 146, 150, 153, 155, 156
MX (Missile Experimental) programme 10, 102–5

Nagasaki 86, 87, 91
National Aeronautics and Space Administration 83
National Institute for Public Policy 66, 67
National Research Council (NRC) 26, 27
National Security Council (NSC) 72, 90
NATO 14, 28, 30, 34, 44, 48, 67, 71, 132, 139, 146, 168, 170, 173, 183, 205, 207, 221, 223, 230, 239
Nature 21
Nautilus 131, 138, 141
Nelkin, D. 38
Nemesis theory 37
New York Times 145
New Zealand 50
Night After: Climatic and Biological Consequences of a Nuclear War 47
Nightingale, Florence 209
Nike-Zeus 108
Nimrod 196, 226
Nitze, Paul 66, 96
Non-Proliferation Treaty 34, 49, 50, 145
NORAD 64
Nordheim 92
Nott, John 208, 209, 210, 212, 222, 226, 228
NSC 68 analysis 66
Nuclear aftermath 15
Nuclear arms race 43, 49, 77, 81
Nuclear arsenals 18, 34
Nuclear bomb 135
Nuclear defence policies 13
Nuclear deterrence
  policies 14, 15
  risks of 15, 44
Nuclear development 129–32
Nuclear Freeze Movement 20, 51, 72, 73

Nuclear links 127
Nuclear policy implications 151–4
Nuclear policy-making 8
Nuclear power 127–66
Nuclear Power Decisions 150
Nuclear stockpiles 30, 42, 29
Nuclear submarines 131, 138, 141, 217
Nuclear war 13
  accidental 14
  climatic consequences of 15, 18–22, 25, 44, 48
  death toll 16
  destructive implications of 17
  potential effects of 14
  probability of 17
  risk of 15, 32
  scenario of 25, 32
  technological hazards 16
Nuclear weapons 8, 14, 23, 34, 127–66
  rendering impotent and obsolete 61–76
Nuclear Weapons: Inquiry, Analysis and Debate 233
Nuclear winter 13–59
  debate outside the USA 43–9
  fears of unsurvivable 31–2
  first strike 31
  global risks posed by 29
  hypothesis of 19–22
  implications outside NATO and Warsaw Pact 49
  notion of 29
  policy implications 22, 25–8
  possibility of 31, 43
  prospect of 15–16
  research implications 15
  risk of 32, 34, 43, 47, 49
  scientific status of hypothesis of 22–5
  Soviet book on 47
  strategic implications of 32
  use of term 14
Nunn, Sam 36, 40
Nye, Joseph 34, 50
Nyerere, Julius 49

Official Secrets Act 214, 233, 234, 240
Oldfield, Sir Maurice 159
O'Neill, Martin 227
Oppenheimer, Robert 87, 91, 92, 94, 240
Owen, David 111

Palme, Olaf 49
Papandreou, Andreas 49
Parliamentary accountability 127
Payne, Keith 30, 66
Pearle, Richard 39
Penny, Sir William 137
Perle, Richard 36, 38, 66
Perry, R. 130
Pershing missiles 32, 44
Philadelphia Inquirer 70
Pike, Summer 94
Pipes, Richard 31, 66
Plowden, Sir Edwin 135, 136
Poland 61
Polaris 78, 109, 146, 149, 217, 218, 219, 231, 232, 238
 A-3 98, 99, 101, 108
 B-3 99
Polaris submarine 109
Policy making 121–5, 167–204
Pollack, James 19–23
Ponting, Clive 211, 235
Poseidon missile 109, 110, 111
Poseidon submarine programme 111
Powell, Sir Richard 136
Pravda 46
Presidential Scientific Advisory Council (PSAC) 83
Pressure groups 65–9
Pressurized water reactor (PWR) 128, 131, 146
Priroda 46, 47
Privileged knowledge 10
Ptarmigan system 194

Quayle, Dan 36

Race to Oblivion: a Participant's View of the Arms Race 78
Radar-jamming techniques 171
Rainbow Warrior 50
RAND Corporation 28
Reagan, Ronald 4, 48, 53, 61–3, 65, 66, 68, 69, 71–4, 145, 146, 194
Reppy, Franklin 232
Research and development (R&D) 4, 6, 10, 11, 77, 80–3, 87, 89, 108, 112, 113, 172, 176–81, 185, 188–90
Reychler, Luc 233
Richtmeyer 92
Rickover, Captain Hyman 130, 131
Rita system 194
Rogers, General Bernard 170

Rostow, Eugene 66
Rotblat, Joseph 18, 216
Rowny, Edward 66
Royal College of Defence Studies 221
Royal Institute of International Affairs 238
Royal Society of Canada 26, 27
Royal United Services Institute 223
Ruddock, Joan 227
Rudney, Robert 233
Rumsfeld, Donald 104
Ryle, Sir Martin 147

Sagan, Carl 19, 21, 22, 27, 28, 30, 31, 36, 37, 40, 41, 42, 52
St Laurent des Eaux gas-graphite plant 154
SALT I Treaty 107
SALT II negotiations 107, 110
Schilling, Warner R. 84
Schlesinger, James 6, 104
Schroeer, Dietrich 80
Schultz, George 65, 71
Science Policy Research Unit 190
Science, Technology and the Nuclear Arms Race 80
Scientific and Technological Resources as Military Assets 3
Scientific policy 127
Scientific pressure groups 86
Scientists Against Nuclear Arms (SANA) 44, 148
Scientist's role in weapons development 77–117
SCOPE (Scientific Committee on Problems of the Environment) 23–8, 38, 44, 45, 48
Second World War 15, 91–3, 125, 129, 152, 181, 210, 216
Secrecy 10, 127
Secretariat of Policy Studies 236
Semiconductor devices 172
Sheet, Robert 64
Shenfield, Steven 46
Sherfield, Lord (formerly Sir Roger Makins) 149
Silicon on sapphire (SOS) 174
Silicon Valley 178
Simpson, John 133, 134, 136, 138, 139, 144
Sizewell B nuclear power station 128, 155

public inquiry   146, 147, 148, 155, 156, 158
SLCM   107
Sloss, Leon   36, 38, 40
Smart, Ian   217, 238
Smart missiles   170
Smith, Chris   159
Smith, Gerard   67
Smoke effects   18–19, 25
Smyth, Henry D.   94
Soviet Academy of Sciences   20, 21, 46
Soviet Union   4, 6, 9, 14, 29–33, 40, 41, 45–9, 51–3, 61, 62, 74, 87, 136, 137
Sputnik   100, 136
Statement on the Defence Estimates   236
Stein   95
Strachan, Hew   221
Strategic Air Command (SAC)   102, 103
Strategic cruise missiles (SCM)   105–8
Strategic Defense Initiative (SDI)   4–5, 7–9, 33, 41, 62, 66, 68, 71, 73, 74, 78, 81, 87, 88, 169, 193, 194, 217
Strauss, Admiral Lewis L.   94, 135, 136, 138
Suitcase bomb   62
System X   198

Telecommunications   182
Teller, Edward   32, 63, 67–9, 71, 79, 86, 90–5
Temperature effects   24, 25, 29
Terrain contour matching (TERCOM) guidance system   106
Texas Instruments   173, 178
Thatcher, Margaret   157, 159, 194, 208
Thermonuclear reactions   91–3
Third World   34, 208
Thomas, Dafydd Elis   157, 227
'Threads' (TV programme)   15
Titan II   99
Toon, Owen   19–23
Tornado multi-role combat aircraft   169
Trawsfynydd   144
Trident   44, 78, 146, 225, 226, 229, 230
  C4   218, 219, 231, 232
  D5   219
Truman, Harry S.   66, 89, 95, 96
Turco, Richard   19–23, 44
Type-23 frigate   170
Type-42 destroyer   169

UK Military R&D   189

Ulam   94
United Kingdom   33, 43–5
United Kingdom Atomic Energy Authority (UKAEA)   129, 135
United States   4, 14, 20, 26, 29, 31–43, 51–3, 61, 62, 145, 146, 206
  administration's reaction   34–5
  Congressional hearings   36–43
United States Atomic Energy Commission   93, 129
United States National Academy of Sciences (NAS)   23
University scientists   82–3
Ustinov, Defence Minister   45
UV-B radiation   24

Van Allen   37
Velikhov, E. P.   20, 22
Velikhov, Yevgeny   45
Very High Speed Integrated Circuit programme   172
Vessey, General John W., Jr.   70
Vietnam War   6

Warner, John W.   36
Warner, Sir Frederick   27
Warsaw Pact   14, 28, 30, 61, 239
Watkins, Admiral James   71
Weapons development, scientist's role in   77–117
Weinberger, Caspar   46, 62, 69
Western European Union (WEU)   206
Westland affair   198–9, 230
Westmoreland, General William   171
Williams, Roger   150
Wilson, Pete   36
Windscale accident   136, 137
Wohlstetter, Albert   85
Wood, Lowell J.   67
Wood Mackenzie   199
Wylfa   150

Yakovlev, A.   48
York, Herbert   78, 81, 85, 97, 113
Younger, George   229

Zircon programme   207
Zuckerman hypothesis   79, 83, 84, 86–90, 92, 96, 97, 103, 105, 106, 111–14, 121
Zuckerman, Lord   77–97, 102, 108, 214, 216

JUN 2 8 1990